ADVANCED METHODS
FOR SOLVING
DIFFERENTIAL EQUATIONS

Staff of Research and Education Association

Research and Education Association
505 Eighth Avenue
New York, N. Y. 10018

ADVANCED METHODS FOR SOLVING
DIFFERENTIAL EQUATIONS

Printed in the United States of America

Library of Congress Catalog Number 82-80750

International Standard Book Number 0-87891-541-9

PREFACE

This book is a useful reference for mathematicians, scientists, and engineers who are interested in increasing their knowledge of differential equations. The book imparts much unity to the subject, and points out the connection between various topics. Emphasis is placed on the practical techniques which can be used to solve equations.

Since the book is intended to be a useful reference source, detailed proofs of theorems are for the most part omitted. However, the conclusions of the theorems are stated in a precise manner, and enough references are given for the benefit of persons interested in finding the steps of the proofs.

Included are unconventional ideas such as the theory of matched asymptotic expansions, which are not, as yet, found in most conventional texts.

The information in this book was originated and sponsored by the National Aeronautics and Space Administration and edited by Marvin E. Goldstein and Willis H. Braun.

CONTENTS

iv

CHAPTER 1

Introduction

In this chapter we shall introduce some preliminary concepts and try to gain a certain amount of insight into the behavior of differential equations and their solutions. Such questions as what is meant by the solution of a differential equation and what is the appropriate number of solutions will be considered with some care.

Leibniz, in 1676, was probably the first to introduce the term "differential equation." However, the study of differential equations had its beginnings somewhat before this time in investigations of physical phenomena. Ever since, developments in the field of differential equations have been closely related to the physical sciences.

1.1 SOLUTION OF EQUATIONS IN GENERAL

In this section we introduce some concepts which are needed for the subsequent presentation. First, in order to obtain a geometric interpretation of certain results, we shall have occasion to interpret a set of n variables, say $x_1, \ldots, x_n$, as coordinate axes in an n-dimensional space. A particular set of values, say $x_1^0, \ldots, x_n^0$, of the n variables are then the coordinates of a point in this space. When $n = 2$ and $n = 3$, these spaces are the very familiar two- and three-dimensional Euclidean spaces. Hence, as will be seen, this procedure allows us to use our geometric intuition about two- and three-dimensional spaces to "picture" results about functions and equations involving any number of variables.

We shall not need to use many of the mathematical properties of the n-dimensional spaces; however, it will be important to have a careful definition of certain special regions in these spaces. Thus, let $x_1^0, \ldots, x_n^0$ be the coordinates [1] of a fixed point in an n-dimensional space, and consider the collection of all points whose coordinates satisfy the inequalities

$$|x_1 - x_1^0| < \delta, \quad . \quad . \quad ., \quad |x_n - x_n^0| < \delta$$

for some positive number δ. When $n = 2$, these points form the interior of a square with center at (x_1^0, x_2^0) and sides of length 2δ. When $n = 3$, the points form the interior of a cube with center at (x_1^0, x_2^0, x_3^0) and sides of length 2δ. The reason for using the word "interior" is that the boundary points which satisfy the *equalities*

$$|x_1 - x_1^0| = \delta, \quad . \quad . \quad ., \quad |x_n - x_n^0| = \delta$$

have been omitted from the collection.

Next we define a *domain D* to be any region of the n-dimensional space which satisfies the following two conditions: (1) there exists a positive number δ for each point $x_1^0, \ldots, x_n^0$ belonging to D such that every point $x_1, \ldots, x_n$ lying in the range

$$|x_1 - x_1^0| < \delta, \quad . \quad . \quad ., \quad |x_n - x_n^0| < \delta$$

is also a point of D; and (2) any two points in this region can be connected by a continuous line. Thus, in a two-dimensional space the first condition requires that we be able to draw around each point of D a square whose interior lies entirely within D. This is illustrated in figure 1–1. The only points where this is not possible are the points which lie on the boundary Γ of D. This is because every square about a point of Γ must include at least some points which do not belong to D. Thus, the first condition serves to exclude the boundary points from D.

The second condition simply requires that the region not be composed of two or more disconnected subregions, as shown for two dimensions in figure 1–2.

A *neighborhood* of the point $x_1^0, \ldots, x_n^0$ is simply a domain which contains this point. Evidently, the collection of all points which satisfy the inequalities in condition 1 form a neighborhood of the point $x_1^0, \ldots, x_n^0$.

---

[1] Instead of saying that $x_1^0, \ldots, x_n^0$ are the coordinates of the point, we shall frequently say that the collection $x_1^0, \ldots, x_n^0$ is the point.

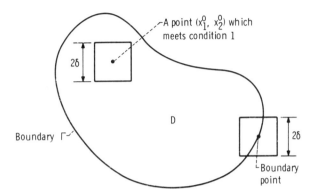

FIGURE 1-1. — Interior and boundary points for a domain D.

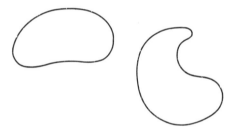

FIGURE 1-2. — Disconnected regions which fail to satisfy condition 2 for a domain.

A *single-valued function* in the x, y-plane can be roughly thought of as one or more curves which associate a single value y with each value x. For example, the function $y = x^2/2$ shown in figure 1-3 associates the single number $x_0^2/2$ with each point x_0. On the other hand, a multivalued function associates more than one value of y with each value of x. For example, the function $\tan^{-1} x$, which is shown in figure 1-4, associates infinitely many values of y with each value of x.

In order to avoid ambiguity *we shall always suppose that every function which is encountered is single valued unless explicitly stated otherwise.*

Consider the equation

$$F(x, y) = 0 \qquad (1-1)$$

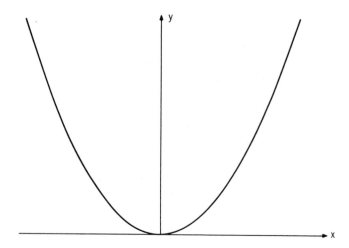

FIGURE 1–3. — Single-valued function $y = x^2/2$.

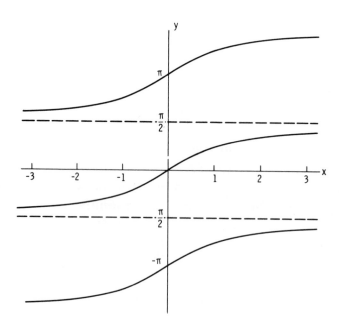

FIGURE 1–4. — Function $y = \tan^{-1} x$.

where F has continuous first partial derivatives with respect to its arguments. We say that the function $y=f(x)$ is a *solution* of this equation if

$$F[x, f(x)] = 0$$

for all values of x for which $f(x)$ is defined. The solutions $y=f(x)$ of equation (1–1) are said to be determined *implicitly* by this equation and are called *implicit functions*.

In fact, the *implicit function theorem* states that for every point (x_0, y_0) such that

$$F(x_0, y_0) = 0$$

and

$$\frac{\partial F}{\partial y}(x_0, y_0) \neq 0 \qquad (1\text{–}2)$$

there is a unique, continuous solution $y=f(x)$ of equation (1–1) which is defined on some neighborhood of the point x_0, which satisfies the condition $y_0=f(x_0)$ and which has a continuous first derivative in this neighborhood.

Now if $F(x, y)$ is a reasonable function, equation (1–1) will be the equation of a curve in the x, y-plane. For example, the implicit relation between x and y

$$F(x, y) \equiv x - y^3 = 0 \qquad (1\text{–}3)$$

is the equation of the cubical parabola shown in figure 1–5. It has an explicit solution given by[2]

$$y = f(x) \equiv (\operatorname{sgn} x) |x|^{1/3} \qquad (1\text{–}4)$$

for all values of x.

---

[2] The function $\operatorname{sgn} x$ is defined by

$$\operatorname{sgn} x = \begin{cases} +1 \text{ for } x \geqslant 0 \\ -1 \text{ for } x < 0 \end{cases}$$

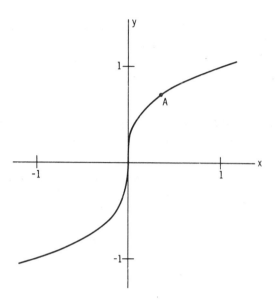

FIGURE 1–5. — Cubical parabola $x = y^3$.

Differentiating this equation with respect to x shows that

$$\frac{dy}{dx} = f'(x) = \frac{1}{3|x|^{2/3}}$$

Hence, condition (1–2) is satisfied at any point A on the curve in figure 1–2 which does not coincide with the origin. And, as required by the theorem, equation (1–4) provides a unique solution with a continuous first derivative in some neighborhood of this point. However, condition (1–2) does not hold at the origin. But then the solution (1–4) has an infinite derivative (which is certainly discontinuous) at this point.

More generally, consider the equation

$$F(x_1, x_2, \ldots, x_n, y) = 0 \qquad (1\text{–}5)$$

A solution to this equation is a function

$$y = f(x_1, x_2, \ldots, x_n)$$

with the property that

$$F[x_1, x_2, \ldots, x_n, f(x_1, x_2, \ldots, x_n)] = 0$$

for all values of $x_1, x_2, \ldots, x_n$ for which f is defined. In order to obtain a geometric picture of equation (1–5) and its solution, it is convenient to think of the variables $x_1, x_2, \ldots, x_n, y$ as the coordinates of a point in an $(n+1)$-dimensional space.

Now suppose that the function F has first partial derivatives with respect to its arguments in some domain D of this space. The implicit function theorem now shows that for every point $(x_1^0, x_2^0, \ldots, x_n^0, y^0)$ of D such that

$$F(x_1^0, x_2^0, \ldots, x_n^0, y^0) = 0$$

$$\frac{\partial F}{\partial y}(x_1^0, x_2^0, \ldots, x_n^0, y^0) \neq 0$$

there is a unique continuous solution

$$y = f(x_1, x_2, \ldots, x_n)$$

of equation (1–5) which is defined in some neighborhood of the point x_1^0, $x_2^0, \ldots, x_n^0$, satisfies the condition $y^0 = f(x_1^0, x_2^0, \ldots, x_n^0)$, *and has continuous first partial derivatives in this neighborhood.*

1.2 DEFINITION OF DIFFERENTIAL EQUATION

A differential equation is an equation connecting the values of a function, called the *dependent variable*, the derivatives of this function, and certain known quantities. If the dependent variable is a function of a single variable (independent) variables, the differential equation is called a *partial differential differential equation*. If the dependent variable is a function of two or more (independent) variables, the differential equation is called a *partial differential equation.*

Thus, an ordinary differential equation is an equation of the form

$$F\left(x, y, \frac{dy}{dx}, \ldots, \frac{d^n y}{dx^n}\right) = 0 \tag{1–6}$$

where the positive integer n, the order of the highest derivative which appears in equation (1–6), is called the *order* of the equation and F is a function of the indicated $n+2$ quantities.

If the function F in equation (1–6) is a polynomial of degree m in the highest order derivative $d^n y/dx^n$, we say that it is a *differential equation of degree m*. Thus, the first-order differential equation of degree m has the form[3]

$$F(x, y, y') \equiv M_0(x, y) (y')^m + \sum_{k=1}^{m} M_k(x, y)(y')^{m-k} = 0$$

The differential equation (1–6) is said to be *linear* if the dependent variable and all its derivatives appear only to the first degree. Thus, the general linear equation can be written in the form

$$a_0(x) \frac{d^n y}{dx^n} + a_1(x) \frac{d^{n-1}y}{dx^{n-1}} + \ldots + a_{n-1}(x) \frac{dy}{dx} + a_n(x)y + b(x) = 0 \quad (1\text{–}7)$$

When the function $b(x)$ in equation (1–7) is identically zero, the equation is said to be *homogeneous*.

1.3 SOLUTIONS AND INTEGRALS OF DIFFERENTIAL EQUATIONS

A *(particular) solution*[4] of the differential equation[5] (1–6) is any n-times differentiable function $f(x)$ defined on some interval $a < x < b$ (which may be infinite) such that equation (1–6) becomes an identity when y and its derivatives are replaced by $f(x)$ and its derivatives. Hence,

$$F\left[x, f(x), \frac{df(x)}{dx}, \ldots, \frac{df^n(x)}{dx^n}\right] = 0$$

for all x in the interval $a < x < b$. Thus, the function $y = \sin x$ is a solution of the differential equation $y'' + y = 0$.

---

[3] We shall frequently write f' or $f'(x)$ for $df(x)/dx$, f'' or $f''(x)$ for d^2f/dx^2, . . ., and $f^{(n)}$ or $f^{(n)}(x)$ in place of $d^n f/dx^n$ for $n = 1, 2, \ldots$

[4] The term "solution" was first used by Lagrange (1774).

[5] There are also solutions known as *weak solutions* which satisfy the differential eq. (1–6) only in a certain average sense. We shall not pursue this topic further here. The interested reader is referred to ref. 1 for an elementary treatment and to ref. 2 for a more advanced discussion.

Now consider the equation

$$F(x, y, c_1, c_2, \ldots, c_n) = 0 \tag{1-8}$$

in which x and y are variables, and $c_1, c_2, \ldots, c_n$ are independent [6] arbitrary constants.[7] We shall suppose that equation (1–8) possesses solutions for at least certain values of the constants $c_1, c_2, \ldots, c_n$.

Upon differentiating equation (1–8) with respect to x, n times in succession, we obtain

$$\frac{\partial F}{\partial x} + \frac{\partial F}{\partial y} y' = 0$$

$$\frac{\partial^2 F}{\partial x^2} + 2 \frac{\partial^2 F}{\partial x \partial y} y' + \frac{\partial^2 F}{\partial y^2} (y')^2 + \frac{\partial F}{\partial y} y'' = 0$$

$$\frac{\partial^n F}{\partial x^n} + \ldots \quad + \ldots \quad + \frac{\partial F}{\partial y} y^{(n)} = 0$$

If the n constants can be eliminated between these n equations and equation (1–8), we can, upon carrying out this elimination, obtain a differential equation [8]

$$G[x, y, y', y'', \ldots, y^{(n)}] = 0 \tag{1-9}$$

It is clear, from the manner in which equation (1–9) was obtained, that every solution of equation (1–8) satisfies[9] equation (1–9). The function F is therefore sometimes referred to as a *primitive* of equation (1–9). In many instances the solutions of equation (1–8) for y as a function of x can be expressed as formulas of the form

---

[6] This means that eq. (1–8) is not expressible in terms of fewer than n constants. For example, $y^2 - x^2 + c_1^2 - c_2$ depends effectively only on the single constant $c = c_1^2 - c_2$.

[7] We assume that all functions are differentiable as many times as is necessary for the argument and that $\partial F/\partial y$ is not identically zero.

[8] In practice it usually will not be possible to carry out all the algebraic operations which are necessary to obtain an explicit formula for eq. (1–9).

[9] However, as we shall see in the following example, the converse of this statement is not necessarily true.

9

$$y = f(x, c_1, c_2, \ldots, c_n) \tag{1-10}$$

which are parameterized by the n constants $c_1, c_2, \ldots, c_n$. We therefore anticipate that, for many nth-order differential equations, there exist formulas of the form (1–10) each of which contains n independent constants $c_1, \ldots, c_n$ and provides a solution to its corresponding equation for every set of values of these n constants for which the formula makes sense.[10] Such a formula, if it exists, is said to be a *general solution* of the equation.

This terminology should not be interpreted to mean, however, that every differential equation possesses a single general solution from which every solution to the equation can be obtained by suitably choosing the values of the constants. This is only true when certain restrictions are imposed on the differential equation (1–6).

For example, consider the equation [11]

$$F(x, y, c) \equiv c^{2-n}[x - c^{(n-1)}]^2 + y^n - 1 = 0 \qquad \text{for } n = 1 \text{ or } n = 2 \tag{1-11}$$

where c is an arbitrary parameter which can take on any real value. By differentiating this equation with respect to x we find that

$$F_x + y'F_y = 2c^{n-1}[x - c^{(n-1)}] + ny^{n-1}y' = 0 \tag{1-12}$$

First, consider the case where $n = 1$. Then upon eliminating c between equations (1–11) and (1–12) we obtain the first-order linear equation

$$y' + \frac{2(1-y)}{x-1} = 0 \tag{1-13}$$

Now, for each value of c, equation (1–11) possesses a single solution

$$y = 1 - c(x-1)^2 \qquad \text{for } -\infty < x < \infty \tag{1-14}$$

This must, therefore, be a general solution to the differential equation (1–13).

---

[10] It is usually necessary to restrict the range of the c's and of x in order to avoid imaginary expressions and other degeneracies.

[11] When $n = 1$, the equation represents a family of parabolas with vertices at the point $x = 1$, $y = 1$; and when $n = 2$, it represents a family of unit circles with centers on the x-axis.

And it happens that every solution to equation (1–13) can be obtained from equation (1–14) by a suitable choice of the constant c.

Next, consider the case where $n=2$. Then upon eliminating c between equations (1–11) and (1–12), we obtain the differential equation

$$y^2(y'^2+1)-1=0 \tag{1–15}$$

For each value of c, equation (1–11) now possesses the two solutions [12]

$$\left.\begin{array}{l} y=\sqrt{1-(x-c)^2} \\ y=-\sqrt{1-(x-c)^2} \end{array}\right\} \quad \text{for } |x-c| \leqslant 1 \tag{1–16}$$

Hence, both of these formulas are general solutions to equation (1–15). Thus, there is no single formula involving only a single parameter from which every solution can be obtained.[13] This is not surprising since equation (1–15) can be written as

$$\frac{1}{4}\left(\frac{dy^2}{dx}\right)^2+y^2-1=0 \tag{1–17}$$

and since only y^2 appears in this equation, it is clear that if $y=f(x)$ is a solution, so is $y=-f(x)$.

Even though equations (1–16) are the only general solutions to equation (1–15), it is still not possible to obtain every solution to the differential equation from these two formulas. Thus, equation (1–15) also possesses the two solutions

$$y=+1$$

$$y=-1$$

which not only cannot be obtained from either of equations (1–16) but do not even satisfy the primitive equation (1–11). They are called *singular* solutions of equation (1–15) (see section 1.5) and are tangent to every solution obtained from the general solutions.

---

[12] For any positive real variable x, $\sqrt{x}$ will always denote the positive square root of x.

[13] Notice that we can combine the two eqs. (1–16) into a single formula $y=(-1)^n \sqrt{1-(x-c)^2}$ for $n=0, 1$, but this solution involves the two parameters n and c.

We have shown how an nth-order differential equation can be obtained by successively differentiating an equation of the type (1–8). In fact if, for a given differential equation, we can find an implicit relation of the type (1–8) such that every solution of the differential equation is also a solution of this equation, we can for practical purposes consider the differential equation solved. Thus, the process of finding a solution to a differential equation might be thought of as reversing the process of obtaining a differential equation from its primitive. The first step of this inverse process when applied to the nth-order differential equation (1–6) (if it can be carried out) leads to a differential equation of order $n-1$ which involves a single arbitrary constant c. And if this equation can be solved for c, it can be written in the form

$$H(x, y, y', \ldots, y^{(n-1)}) = c \tag{1-18}$$

More generally, if for any given function H every solution of equation (1–6) satisfies an equation of the form (1–18) for some value of c, this function (and sometimes eq. (1–18) itself) is called a *(first) integral* of equation [14] (1–6). Geometrically, this means that an integral H is constant along every solution curve $y = f(x)$ of the differential equation.

1.4 SOLUTIONS TO NORMAL EQUATIONS

If we try to find a solution to equation (1–15), say

$$y = f(x)$$

which satisfies the initial condition $f(1) = 0$, we find from the first equation (1–16) that

$$y = f(x) = \sqrt{1 - x^2}$$

and from the second equation (1–16) that

$$y = f(x) = -\sqrt{1 - x^2}$$

---

[14]Since integration is the inverse of differentiation, the preceding remarks show why the process of solving a differential equation is sometimes referred to as integrating the equation.

Therefore, equation (1–15) has two solutions which satisfy the condition $y=0$ at $x=1$. However, we usually expect the solution of any first-order differential equation arising from a physical problem to be uniquely determined by a single initial condition. It is, therefore, necessary to find what conditions must be imposed on a given differential equation if our expectations are to be justified. To this end suppose that equation (1–6) can be solved for its highest order derivative. It can then be written as an equation of the form

$$y^{(n)} = G(x, y, y', \ldots, y^{(n-1)}) \tag{1–19}$$

Any ordinary differential equation which has the form of equation (1–19) is called a *normal* differential equation or is said to be in *normal form*.

We are interested in finding conditions which ensure that equations of this type possess unique solutions satisfying the *initial conditions*

$$x = x_0, y = Y_0, y' = Y_1, \ldots, y^{(n-1)} = Y_{n-1} \tag{1–20}$$

where $Y_0, \ldots, Y_{n-1}$ are constants. *The fundamental theorem* for nth-order differential equations states that this is always the case, at least locally, provided that certain restrictions are imposed on the function G. In order to state this theorem in a precise way, let us for the moment treat the variables $x, y, y', \ldots, y^{(n-1)}$ as independent. Then if there exists a positive number δ such that the functions

$$G, \frac{\partial G}{\partial y}, \frac{\partial^2 G}{\partial y^2}, \ldots, \frac{\partial^{n-1} G}{\partial y^{n-1}} \tag{1–21}$$

are defined and continuous[15] for all values of the variables $x, y, y', \ldots, y^{(n-1)}$ which lie in the range

$$|x - x_0| < \delta, |y - y_0| < \delta, |y' - Y_1| < \delta, \ldots, |y^{(n-1)} - Y_{n-1}| < \delta$$

the fundamental theorem states that equation (1–19) possesses a solution $y = f(x)$ which satisfies the initial conditions (1–20) and is defined on some

---

[15] Notice that only the partial derivatives of G with respect to the variables $y, y', \ldots, y^{(n)}$ need be continuous, but the partial derivative with respect to x need not even exist.

13

interval containing the point $x = x_0$, say

$$a < x < b \qquad (1\text{--}22)$$

In addition, this solution is unique. This means that if $g(x)$ is any other solution to equation (1–19) which satisfies the initial conditions (1–20), then

$$f(x) \equiv g(x)$$

for all x in the interval[16] (1–22).

Notice that we have only asserted that the interval (1–22) exists; we have not given any idea of its size.[17] That is, there is no relation given between a, b, and δ. The interval might be exceedingly small. However, we shall give a measure of the size of this interval for an important special case. Thus, if the partial derivatives (1–21) are defined and continuous for all values of y, y', . . ., $y^{(n-1)}$ and for all values of x in some interval which contains the point $x = x_0$, say

$$\alpha < x < \beta \qquad (1\text{--}23)$$

and if there exists a single constant K such that

$$\frac{\partial G}{\partial y^{(i)}} \leq K \quad \text{for } i = 0, 1, 2, \ldots, n-1$$

for all y, . . ., $y^{(n-1)}$ and all x in the interval (1–23), the solution $y = f(x)$ which satisfies the initial conditions (1–20) exists and satisfies the differential equation for *all* x *on the interval (1–23)*.

For example, it is easy to verify that the equation

$$y' = G(y) \equiv 1 + y^2$$

has the solution $y = \tan(x + c)$. Since $\tan x$ becomes infinite at $x = \pm\pi/2$, it

---

[16] Actually, the same conclusions can be reached even when somewhat weaker conditions are imposed on the function G in eq. (1–19). Proofs of the fundamental theorem can be found in refs. 3 to 5.

[17] In fact, a measure of the size of this interval can be given in terms of certain bounds on the function G. See, for example, ref. 4, theorem 8, p. 118.

is clear that the solutions to this differential equation are defined only on intervals whose lengths are at most π, even though G and $\partial G/\partial y$ are defined and continuous everywhere. This occurs because there is no constant K which is larger than $\partial G/\partial y = 2y$ for all values of y.

Although we normally expect the solution of an nth-order differential equation to be uniquely determined by specifying n initial conditions (or the equivalent), we have seen by example that specifying these conditions is not necessarily sufficient to uniquely determine the solution to every nth-order equation. The preceding discussion indicates that the extraneous solutions must arise either because the differential equation (1–6) is not in normal form (i.e., solved in a unique way for the highest order derivative) or because, even if it is in the normal form (1–19), the function G is not sufficiently smooth at some points.

A general solution to equation (1–19), if it exists, contains n "arbitrary" constants. And since, in principle, it is usually possible to solve n equations in n unknowns, this is consistent with the fundamental theorem which states that n initial conditions determine the solution to equation (1–19).

For example, integrating the differential equation

$$y'' = x^2 \tag{1-24}$$

twice yields the general solution

$$y = \frac{x^4}{12} + c_1 x + c_2 \tag{1-25}$$

containing two arbitrary constants c_1 and c_2. The constants are uniquely determined by the initial conditions

$$x = x_0, y = Y_0, y' = Y_1$$

since the two equations

$$Y_0 = \frac{x_0^4}{12} + c_1 x_0 + c_2$$

$$Y_1 = \frac{x_0^3}{3} + c_1$$

15

can be solved for c_1 and c_2 to obtain

$$c_1 = Y_1 - \frac{x_0^3}{3}$$

$$c_2 = Y_0 - x_0 Y_1 + \frac{x_0^4}{4}$$

More generally, suppose that by some formal process of integration we have found m general solutions

$$y = f_i(x, c_1, c_2, \ldots, c_n) \qquad i = 1, 2, \ldots, m \tag{1-26}$$

of the differential equation (1–19) and that every solution of this differential equation can be obtained from these formulas by properly choosing the constants. Then for every set of values of the n constants $x_0, Y_0, Y_1, \ldots, Y_{n-1}$ for which G satisfies the conditions imposed in the fundamental theorem, there must be one and only one value of i for which it is possible to solve the n equations

$$Y_0 = f_i(x_0, c_1, c_2, \ldots, c_n)$$

$$Y_1 = \frac{df_i}{dx}(x_0, c_1, c_2, \ldots, c_n)$$

$$\vdots$$

$$Y_{n-1} = \frac{d^{n-1}f_i}{dx^{n-1}}(x_0, c_1, c_2, \ldots, c_n)$$

for the n constants $c_1, c_2, \ldots, c_n$.

For example, the function

$$y = f(x, c) \equiv \sqrt{1 + ce^x} \tag{1-27}$$

satisfies the differential equation

$$y' = F(y) \equiv \frac{1}{2}\left(y - \frac{1}{y}\right) \qquad (1\text{--}28)$$

for all values of c for which the radical exists. And for this equation the partial derivatives (1–21) are always continuous except when $y=0$. Hence, the fundamental theorem shows that there must be a solution which satisfies the initial condition

$$x=0 \qquad y=-1$$

However, there is no possible choice of c in equation (1–27) for which $f(0, c) = -1$, which shows that it is not possible to obtain every solution to equation (1–28) from the formula (1–27). However, the function

$$y = -\sqrt{1 + ce^x} \qquad (1\text{--}29)$$

also satisfies equation (1–28) for all values of c for which the radical exists. Equations (1–27) and (1–29) taken together will now provide a unique solution for each initial condition which does not involve $y=0$. But these formulas will provide two solutions satisfying each initial condition involving $y=0$, where the partial derivatives (1–21) are not continuous. However, the fundamental theorem provides no information about solutions which pass through points where the partial derivatives (1–21) are not continuous.

Specifying initial conditions is only one of many ways of determining the values of the arbitrary constants which appear in the general solutions of a differential equation. The most common alternative is to require that the solution and its derivatives satisfy certain conditions at both ends of some interval, say $x_1 \le x \le x_2$, within which this solution is being sought. These conditions are called *boundary conditions*.[18]

For example, we might require that the solution to equation (1–24) satisfy the boundary conditions

---

[18] The term "initial conditions" arose in conjunction with problems in mechanics which have time as the independent variable and have conditions imposed at some initial time. The term "boundary conditions" arose in conjunction with problems involving physical distance as the independent variable with conditions imposed at the boundaries of a physical region.

$$y = y_1 \qquad \text{at } x = x_1$$

$$y = y_2 \qquad \text{at } x = x_2$$

Then substituting equation (1–25) into these conditions shows that

$$y_1 = \frac{x_1^4}{12} + c_1 x_1 + c_2$$

$$y_2 = \frac{x_2^4}{12} + c_1 x_2 + c_2$$

And since these equations have the unique solution

$$c_1 = \frac{y_1 - y_2}{x_1 - x_2} + \frac{x_2^4 - x_1^4}{12(x_1 - x_2)}$$

$$c_2 = \frac{x_2 y_1 - x_1 y_2}{x_2 - x_1} + \frac{x_1 x_2 (x_2^3 - x_1^3)}{12(x_2 - x_1)}$$

we conclude that the two boundary conditions uniquely determine the solution of equation (1–24). There are a number of theorems which give conditions which, if satisfied, will ensure that the solution of an nth-order equation will be uniquely determined by n boundary conditions. However, since these conditions are fairly complicated, we will not state any of these theorems herein.

1.5 SOLUTIONS TO EQUATIONS NOT IN NORMAL FORM—SINGULAR SOLUTIONS

We have shown that the nth-order normal differential equation possesses a solution which is uniquely determined by n initial conditions, provided that the conditions imposed in the fundamental theorem are satisfied. In order to use this result to obtain information about the solutions of a differential equation which is not in normal form, it is necessary to solve this equation, at least in principle, for its highest order derivative.

First, consider the general nth-order equation of the first degree

$$U(x, y, \ldots, y^{(n-1)}) y^{(n)} + V(x, y, \ldots, y^{(n-1)}) = 0 \qquad (1\text{--}30)$$

This equation can be written in the normal form

$$y^{(n)} = G(x, y, \ldots, y^{(n-1)}) \equiv -\frac{V(x, y, \ldots, y^{(n-1)})}{U(x, y, \ldots, y^{(n-1)})} \qquad (1\text{-}31)$$

for all values of $x, y, \ldots, y^{(n-1)}$ for which U is not equal to zero. We shall suppose that U and V are defined and possess continuous first partial derivatives for all values of $x, y, \ldots, y^{(n-1)}$. Then the function G will possess continuous first partial derivatives for all values of $x, y, \ldots, y^{(n-1)}$ at which $U \neq 0$. Hence, the fundamental theorem shows that equation (1-30) will possess a unique solution satisfying any set of n initial conditions $x = x_0, y = y_0, \ldots, y^{(n-1)} = Y_{n-1}$ for which $U(x_0, Y_0, \ldots, Y_{n-1}) \neq 0$. Each solution of equation (1-31) will satisfy equation (1-30); and if U is never equal to zero, every solution of equation (1-30) will satisfy equation (1-31). Hence, in this case, equations (1-30) and (1-31) are equivalent. However, suppose that the function U vanishes for certain values of its arguments. If equation (1-30) possesses a solution $y = f(x)$ such that for certain values[19] of x

$$U(x, f(x), \ldots, f^{(n-1)}(x)) = 0 \qquad (1\text{-}32)$$

then $y = f(x)$ will not necessarily satisfy equation (1-31) for these values[20] of x. It is called a *singular solution* of equation (1-30).

For example, the first-order equation

$$yy' - (y^2 - y) = 0 \qquad (1\text{-}33)$$

can be written in the normal form

$$y' = \frac{y^2 - y}{y} = y - 1 \qquad (1\text{-}34)$$

The normal equation (1-34) has the general solution

$$y = 1 + ce^x$$

---

[19] Thus $y = f(x)$ satisfies both eq. (1-30) and the $(n-1)$st-order differential equation $U(x, y, \ldots, y^{(n-1)}) = 0$.

[20] Even if the numerator of G vanishes in such a way that it is possible to define $G(x, f(x), \ldots, f^{(n-1)}(x))$ at these values of x.

and this is also a general solution of equation (1-33). However, equation (1-33) also has the solution

$$y = 0 \tag{1-35}$$

but equation (1-34) does not. Since the coefficient of y' vanishes at $y = 0$, equation (1-35) is a singular solution of equation (1-33).

Now consider the general equation (1-6) and let us temporarily treat the quantities $x, y, \ldots, y^{(n)}$ as independent variables. Suppose that F has continuous first partial derivatives with respect to its arguments. Then the implicit function theorem (given in section 1.1) shows that equation (1-6) has several (possibly infinitely many) solutions of the form

$$y^{(n)} = G(x, y, \ldots, y^{(n-1)}) \tag{1-36}$$

where the function G is continuous and has continuous first partial derivatives for all values of its arguments for which it is defined. One of these solutions will satisfy equation (1-6) for each set of values of $x, y, \ldots, y^{(n)}$ for which equation (1-6) holds and

$$\frac{\partial F}{\partial y^{(n)}}(x, y, \ldots, y^{(n)}) \neq 0 \tag{1-37}$$

However, equation (1-36) is a normal differential equation; and therefore the fundamental theorem shows that it possesses a unique solution satisfying any set of initial conditions $x = x_0, y = Y_0, \ldots, y^{(n-1)} = Y_{n-1}$ for which $G(x_0, Y_0, \ldots, Y_{n-1})$ is defined. All the solutions of the normal equations of the form (1-36) obtained from equation (1-6) will satisfy equation (1-6). If $\partial F / \partial y^{(n)}$ is never equal to zero for any values of $x, y, \ldots, y^{(n)}$ for which equation (1-6) holds, every solution of equation (1-6) will satisfy one of the equations (1-36). Hence, in this case, equation (1-6) is equivalent to the set of normal equations of the form (1-36). However, suppose that

$$\frac{\partial F}{\partial y^{(n)}}(x, y, y', \ldots, y^{(n)}) = 0 \tag{1-38}$$

for certain values of $x, y, \ldots, y^{(n)}$. If equation (1-6) possesses a solution

$y = f(x)$ such that for certain values [21] of x

$$\frac{\partial F}{\partial y^{(n)}} (x, f(x), \ldots, f^{(n)}(x)) = 0 \tag{1-39}$$

then for these values of x the implicit function theorem no longer guarantees that there will be normal equations (satisfying the conditions imposed in the fundamental theorem) which are satisfied by the solution $y = f(x)$ of equation (1–6). This solution is, therefore, called a *singular solution* of equation (1–6).

For example, the equation

$$F(y, y') \equiv y'^2 + 1 - y = 0 \tag{1-40}$$

has the two solutions

$$y' = G_1(y) \equiv \sqrt{y-1} \tag{1-41}$$

$$y' = G_2(y) \equiv -\sqrt{y-1} \tag{1-42}$$

where G_1 and G_2 have continuous derivatives for all values y, y' for which equation (1–40) holds and for which

$$\frac{\partial F}{\partial y'} = 2y' \neq 0 \tag{1-43}$$

Now equation (1–41) has the solution

$$y = 1 + \frac{1}{4}(x-c)^2 \qquad \text{for } x \geqslant c \tag{1-44}$$

and equation (1–42) has the solution

$$y = 1 + \frac{1}{4}(x-c)^2 \qquad \text{for } x \leqslant c \tag{1-45}$$

---

[21] Thus $y = f(x)$ satisfies both the differential eq. (1–6) and the differential eq. (1–38).

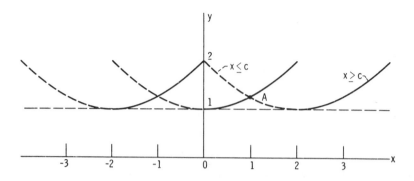

FIGURE 1-6. — Solutions of equations (1-41) and (1-42).

For each value of c, these two solutions are the two branches of the same parabola. They are shown in figure 1-6. The solutions of equation (1-41) are shown as solid curves. The solutions of equation (1-42) are shown dashed. Notice that one solution to equation (1-41) and one solution to equation (1-42) pass through each point (such as point A) lying above the line $y=1$, which is consistent with the fundamental theorem.

Equation (1-43) shows that the points where $y'=0$ are exceptional. But, for those values of y and y' which satisfy equation (1-40), this can occur only when $y=1$. However, when $y=1$, dG_1/dy and dG_2/dy become infinite; and therefore the fundamental theorem does not apply along this line. Equations (1-41) and (1-42) (and, therefore, also eq. (1-40)) possess the solution

$$y=1$$

which lies along this line and which cannot be obtained for any choice of c from either equation (1-44) or (1-45). Thus, $y=1$ is a singular solution. Notice that there are three solutions of equation (1-40) passing through each point on the line $y=1$.

Now consider the first-order differential equation

$$F(x, y, y') = 0 \qquad (1-46)$$

The singular solutions of this equation, if they exist, must also satisfy the equation

$$\frac{\partial F}{\partial y'}(x, y, y') = 0 \tag{1-47}$$

When y' can be eliminated between equations (1–40) and (1–41), we obtain the equation

$$g(x, y) = 0 \tag{1-48}$$

Since the singular solutions simultaneously satisfy equations (1–46) and (1–47), they must also satisfy equation (1–48). Thus, every singular solution of equation (1–46) can be found by solving equation (1–48). However, the converse is by no means always true. Since the symbol p is often used to denote y', equation (1–48) is sometimes called the *p-discriminant* equation and the curve described by this equation is sometimes called the *singular locus*. When attempting to find all solutions of a first-order equation, the solutions to the *p*-discriminant equation should be checked to see if they also satisfy the differential equation.

For example, consider the first-order quadratic equation

$$F(x, y, y') = Ay'^2 + By' + C = 0$$

where A, B, and C are functions of x and y. Upon eliminating y' between this equation and the equation

$$\frac{\partial F}{\partial y'} = 2Ay' + B = 0$$

we find that the *p*-discriminant equation is

$$B^2 - 4AC = 0$$

Thus, for the first-order equation

$$F(x, y, y') \equiv xy'^2 - 3yy' + 9x^2 = 0 \tag{1-49}$$

the *p*-discriminant equation is

$$y^2 = 4x^3$$

which has the two solutions $y = \pm 2x^{3/2}$. Direct substitution shows that both these solutions satisfy equation (1–49); and, therefore, they are both singular solutions.

1.6 LINEAR EQUATIONS

In this section we present a number of important special properties which are possessed by the solutions of linear equations. The general form of the nth-order linear equation is given in equation (1–7). If $f_1(x)$ and $f_2(x)$ are any two solutions to a linear homogeneous differential equation and c_1 and c_2 are any constants, the function

$$y = c_1 f_1(x) + c_2 f_2(x) \tag{1-50}$$

is also a solution. This linear *superposition principle* is extremely important in analysis. It can easily be extended to include linear combinations of any number of solutions. Notice that the homogeneous equation always possesses the *trivial* solution $y = 0$.

The equation obtained from a particular linear equation of the form (1–7) by setting $b(x)$ equal to zero is called the *associated homogeneous equation* of this equation. Any solution to the associated homogeneous equation is called a *homogeneous solution* of the equation. *If* f(x) *is any solution of equation* (1–7) *and if* f$_H$(x) *is any homogeneous solution of this equation, the function*

$$f(x) + cf_H(x)$$

is also a solution. Of course, the superposition principle holds only for the homogeneous solutions of a linear equation.

Notice that the linear equation (1–7) can always be written in the normal form (1–19) with

$$G = -\frac{a_1}{a_0} y^{(n-1)} - \ldots - \frac{a_{n-1}}{a_0} y' - \frac{a_n}{a_0} y - b \tag{1-51}$$

for all values of x where the coefficient $a_0(x)$ is not equal to zero. It follows that for these values of x

$$\frac{\partial G}{\partial y^{(j)}} = -\frac{a_{n-j}(x)}{a_0(x)} \qquad \text{for } j = 0, 1, 2, \ldots, n-1 \tag{1-52}$$

Now suppose that the coefficients $a_0, a_1, \ldots, a_n, b$ of equation (1–7) are all continuous in some *finite* [22] interval

$$\alpha \leqslant x \leqslant \beta \qquad\qquad (1\text{–}53)$$

Then if $a_0(x)$ does not vanish at any point of this interval, it will always be possible to find a constant K such that

$$-\frac{a_{n-j}(x)}{a_0(x)} \leqslant K \qquad \text{for } j=1,2,3,\ldots,n-1$$

and for all x in the interval (1–53). It can also be seen from equations (1–51) and (1–52) that the partial derivatives (1–21) are continuous for all values of $y, y', \ldots, y^{(n-1)}$ and all values of x in this interval. Hence, in this case, G not only satisfies the conditions imposed in the fundamental theorem, but it satisfies the more restrictive conditions given immediately following it.

We therefore conclude that for linear equations the fundamental theorem can be stated in the following way:

Let the coefficients $a_0(x), a_1(x), \ldots, a_n(x), b(x)$ *of equation (1–7) be continuous on some interval* $\alpha < x < \beta$ *which contains the point* x_0, *and suppose that the function* $a_0(x)$ *does not vanish in this interval. Then for any real numbers* $Y_0, \ldots, Y_n$ *there is a unique solution* $y = f(x)$ *of equation (1–7) satisfying the initial conditions*

$$y(x_0)=Y_0,\, y'(x_0)=Y_1,\, \ldots,\, y^{(n-1)}(x_0)=Y_{n-1}$$

And this solution satisfies the differential equation on the entire interval $\alpha < x < \beta$.

If the linear equation (1–7) has coefficients which are continuous in some interval, the points in that interval where the coefficient $a_0(x)$ vanishes are called *singular points* of the equation. We shall have more to say about such points subsequently.

A set of functions $g_1(x), g_2(x), \ldots, g_n(x)$ is said to be *linearly independent* in an interval if it is impossible to find constants $c_1, \ldots, c_n$ which are not all zeros such that the expression

---

[22] This means that neither α nor β is equal to infinity.

$$c_1 g_1(x) + c_2 g_2(x) + \ldots + c_n g_n(x)$$

is equal to zero at all points (i.e., identically) of the interval. A set of functions which is *not* linearly independent is said to be *linearly dependent*.

If the set of functions $g_1(x), \ldots, g_n(x)$ is linearly independent, no one of these functions can be expressed as a linear combination of the others. Hence, if $a_i \neq 0$, it will not be possible to express the function

$$f(x) = a_1 g_1(x) + a_2 g_2(x) + \ldots + a_n g_n(x) \tag{1-54}$$

in the form

$$f(x) = e_1 g_1(x) + e_2 g_2(x) + \ldots + e_{i-1} g_{i-1}(x) + e_{i+1} g_{i+1}(x) + \ldots + e_n g_n(x)$$

in which the function $g_i(x)$ no longer appears. Thus the constants appearing in equation (1-54) are independent.[23]

For example, the functions $g_1(x) = 2x - 5$ and $g_2(x) = 6x - 15$ are not linearly independent since

$$c_1 g_1(x) + c_2 g_2(x) = 0$$

when $c_1 = 3$ and $c_2 = -1$. However, the functions

$$g_1(x) = 2x - 5 \quad \text{and} \quad g_2(x) = 2x + 5$$

are linearly independent.

Let $g_1, g_2, \ldots, g_n$ be a set of $(n-1)$-times continuously differentiable functions. The determinant

$$W(g_1, g_2, \ldots, g_n) \equiv \begin{vmatrix} g_1 & g_2 & \cdots & g_n \\ g_1' & g_2' & \cdots & g_n' \\ \cdot & & & \\ \cdot & & & \\ \cdot & & & \\ g_1^{(n-1)} & g_2^{(n-1)} & \cdots & g_n^{(n-1)} \end{vmatrix}$$

---

[23] This means that eq. (1-54) is not expressible in terms of fewer than n constants.

is called the *Wronskian* of these functions. Now suppose that the functions $g_1, g_2, \ldots, g_n$ are linearly dependent on some interval. Then there exist constants $c_1, c_2, \ldots, c_n$ not all zero such that

$$c_1 g_1(x) + c_2 g_2(x) + \ldots + c_n g_n(x) = 0$$

at every point of the interval. Hence, we can differentiate this expression $n - 1$ times to obtain

$$c_1 g_1'(x) \quad + \quad c_2 g_2'(x) \quad + \ldots + \quad c_n g_n'(x) \quad = 0$$

$$\cdot \qquad\qquad \cdot \qquad\qquad\qquad \cdot$$
$$\cdot \qquad\qquad \cdot \qquad\qquad\qquad \cdot$$
$$\cdot \qquad\qquad \cdot \qquad\qquad\qquad \cdot$$

$$c_1 g_1^{(n-1)}(x) + c_2 g_2^{(n-1)}(x) + \ldots + c_n g_n^{(n-1)}(x) = 0$$

This gives us a set of n linear equations for the n nonzero constants $c_1, c_2, \ldots, c_n$. If these equations are to be solvable for the constants at all points of the interval, the Wronskian must vanish at all points of this interval. Hence, we can conclude that *if the set* $g_1, g_2, \ldots, g_n$ *of* $(n - 1)$-*times continuously differentiable functions is linearly dependent on the interval* $\alpha \leqslant x \leqslant \beta$, *the Wronskian of these functions must vanish at all points of this interval.*

Another way of saying this is that *if the Wronskian of a set* $g_1, g_2, \ldots, g_n$ *of* $(n - 1)$-*times continuously differentiable functions does not vanish at every point of the interval* $\alpha \leqslant x \leqslant \beta$, *these functions are linearly independent on this interval.*

Thus, the Wronskian of the functions (discussed in the previous example) $g_1(x) = 2x - 5$ and $g_2(x) = 6x - 15$ is

$$W(g_1, g_2) = \begin{vmatrix} 2x - 5 & 6x - 15 \\ 2 & 6 \end{vmatrix}$$

$$= (2x - 5)6 - 2(6x - 15) \equiv 0$$

And this shows that these functions are linearly dependent. However, the Wronskian of the linearly independent functions $g_1(x) = 2x - 5$ and $g_2(x) = 2x + 5$ is

$$W(g_1, g_2) = \begin{vmatrix} 2x-5 & 2x+5 \\ 2 & 2 \end{vmatrix} = -20$$

which is certainly not equal to zero.

Now consider the associated homogeneous equation of equation (1–7)

$$a_0 y^{(n)} + a_1 y^{(n-1)} + \ldots + a_{n-1} y' + a_n y = 0 \qquad (1\text{–}55)$$

and suppose that the coefficients $a_0, a_1, \ldots, a_n$ are continuous in the interval

$$\alpha \leqslant x \leqslant \beta \qquad (1\text{–}56)$$

and that $a_0(x)$ does not vanish on this interval. We shall now show that *any* n *solutions* $f_1(x), \ldots, f_n(x)$ *of equation (1–55) are linearly dependent on the interval (1–56) if, and only if, their Wronskian* $W(f_1, f_2, \ldots, f_n)$ *vanishes at every point of this interval.*

Another way of saying this is that *the* n *solutions* $f_1(x), \ldots, f_n(x)$ *of equation (1–55) are linearly independent on the interval (1–56) if, and only if, the Wronskian* $W(f_1, \ldots, f_n)$ *is not equal to zero at some point of this interval.*

We have already shown that any set of $(n-1)$-times continuously differentiable functions is linearly dependent on an interval only if their Wronskian vanishes at every point of this interval. Hence, it is certainly true for a set of n solutions.[24] In order to show that, conversely, the vanishing of the Wronskian implies that the solutions are linearly dependent, suppose that $W(f_1, f_2, \ldots, f_n)$ vanishes at every point of the interval (1–56) and let x_0 be any point of this interval. Then the Wronskian certainly vanishes at this point, and therefore the system of n equations

---

[24] Recall that the definition of a solution requires that it be $n-1$ times continuously differentiable.

$$\left.\begin{array}{l} c_1 f_1(x_0) \;+\; c_2 f_2(x_0) \;+\; \ldots \;+\; c_n f_n(x_0) = 0 \\[4pt] c_1 f_1'(x_0) \;+\; c_2 f_2'(x_0) \;+\; \ldots \;+\; c_n f_n'(x_0) = 0 \\[4pt] \qquad\cdot\qquad\qquad\cdot\qquad\qquad\qquad\cdot \\[2pt] \qquad\quad\cdot\qquad\qquad\cdot\qquad\qquad\qquad\cdot \\[2pt] \qquad\quad\cdot\qquad\qquad\cdot\qquad\qquad\qquad\cdot \\[4pt] c_1 f_1^{(n-1)}(x_0) \;+\; c_2 f_2^{(n-1)}(x_0) \;+\; \ldots \;+\; c_n f_n^{(n-1)}(x_0) = 0 \end{array}\right\} \qquad (1\text{--}57)$$

has a solution $\tilde{c}_1 = c_1,\ \tilde{c}_2 = c_2,\ \ldots,\ \tilde{c}_n = c_n$ such that the constants $\tilde{c}_1, \tilde{c}_2, \ldots, \tilde{c}_n$ are not all zeros. Hence, we can define a function $f(x)$ by

$$f(x) = \tilde{c}_1 f_1(x) + \tilde{c}_2 f_2(x) + \ldots + \tilde{c}_n f_n(x) \qquad (1\text{--}58)$$

Then it follows from the linear superposition principle that $f(x)$ is a solution to equation (1–55) in the interval (1–56). And differentiating equation (1–58) $n-1$ times shows that

$$f = \tilde{c}_1 f_1 \;+\; \tilde{c}_2 f_2 \;+\; \ldots \;+\; \tilde{c}_n f_n$$

$$f' = \tilde{c}_1 f_1' \;+\; \tilde{c}_2 f_2' \;+\; \ldots \;+\; \tilde{c}_n f_n'$$

$$\cdot\qquad\cdot\qquad\cdot\qquad\qquad\cdot$$
$$\cdot\qquad\cdot\qquad\cdot\qquad\qquad\cdot$$
$$\cdot\qquad\cdot\qquad\cdot\qquad\qquad\cdot$$

$$f^{(n-1)} = \tilde{e}_1 f_1^{(n-1)} - \tilde{c}_2 f_2^{(n-1)} + \ldots + \tilde{c}_n f_n^{(n-1)}$$

Hence, upon setting $x = x_0$ in this system and using equation (1–57), we find that the solution f satisfies the n initial conditions

$$f(x_0) = f'(x_0) = \ldots = f^{(n-1)}(x_0) = 0$$

Since the trivial solution of the homogeneous equation (1–55) also satisfies these conditions and the fundamental theorem shows that there is only one such solution, we therefore conclude that $f(x)$ is equal to zero at every point of the interval (1–56). Therefore,

$$\tilde{c}_1 f_1(x) + \tilde{c}_2 f_2(x) + \ldots + \tilde{c}_n f_n(x) = 0$$

at every point of the interval (1–56). But since the constants $\tilde{c}_1$, $\tilde{c}_2$, . . ., $\tilde{c}_n$ are not all zero, this implies that the set $f_1, f_2, \ldots, f_n$ is linearly dependent and therefore proves the assertion.

We shall now show that *the linear homogeneous equation (1–55) always has* n *linearly independent solutions on the interval (1–56)*. To this end notice that for any point x_0 of this interval the fundamental theorem shows that equation (1–55) has n solutions $f_1(x), f_2(x), \ldots, f_n(x)$ which satisfy the initial conditions

$$f_1(x_0) = 1$$

$$f_1'(x_0) = f_1''(x_0) = \ldots = f_1^{(n-1)}(x_0) = 0$$

and for $r = 2, 3, \ldots, n$

$$f_r(x_0) = f_r'(x_0) = \ldots = f_r^{(r-2)}(x_0) = 0$$

$$f_r^{(r-1)}(x_0) = 1$$

$$f_r^{(r)}(x_0) = f_r^{(r+1)}(x_0) = \ldots = f_r^{(n-1)}(x_0) = 0$$

It is easy to see from these conditions that the Wronskian of these n solutions has the value unity at the point x_0. Thus, it is not zero at every point of the interval $\alpha \leqslant x \leqslant \beta$, and we can therefore conclude that these n solutions are linearly independent.

For example, the second-order differential equation

$$\frac{d^2 y}{dx^2} + y = 0$$

has the two solutions $y = \sin x$ and $y = \cos x$.

The Wronskian of these solutions is

$$W(\sin x, \cos x) = \begin{vmatrix} \sin x & \cos x \\ \cos x & \sin x \end{vmatrix}$$

$$= -\sin^2 x - \cos^2 x = -1 \neq 0$$

Hence, the solutions are linearly independent.

If f_1, f_2, . . ., f_n are any n linearly independent solutions of equation (1–55) on the interval (1–56), and if h(x) is any other solution of equation (1–55) on this interval, there exist constants $\tilde{c}_1$, $\tilde{c}_2$, . . ., $\tilde{c}_n$ such that

$$h(x) = \tilde{c}_1 f_1(x) + \tilde{c}_2 f_2(x) + \ . \ . \ . + \tilde{c}_n f_n(x) \qquad (1-59)$$

on the entire interval.

This means that every solution to the equation on this interval can be obtained from the general solution

$$f(x, c_1, c_2, \ . \ . \ ., c_n) = c_1 f_1(x) + c_2 f_2(x) + \ . \ . \ . + c_n f_n(x)$$

In order to prove this assertion, notice that since f_1, . . ., f_n are linearly independent, $W(f_1, f_2, \ldots, f_n)$ cannot be zero at every point of the interval (1–56). Hence, there exists a point, say x_0, for which the Wronskian is not zero. The n equations

$$
\left.
\begin{aligned}
h(x_0) &= \tilde{c}_1 f_1(x_0) \quad + \quad \tilde{c}_2 f_2(x_0) + \ . \ . \ . + \tilde{c}_n f_n(x_0) \\
h'(x_0) &= \tilde{c}_1 f_1'(x_0) \quad + \quad \tilde{c}_2 f_2'(x_0) + \ . \ . \ . + \tilde{c}_n f_n'(x_0) \\
&\quad\ \cdot \qquad\quad\ \cdot \qquad\quad\ \cdot \qquad\qquad\quad \cdot \\
&\quad\ \cdot \qquad\quad\ \cdot \qquad\quad\ \cdot \qquad\qquad\quad \cdot \\
&\quad\ \cdot \qquad\quad\ \cdot \qquad\quad\ \cdot \qquad\qquad\quad \cdot \\
h^{(n-1)}(x_0) &= \tilde{c}_1 f_1^{(n-1)}(x_0) + \tilde{c}_2 f_2^{(n-1)}(x_0) + \ . \ . \ . + \tilde{c} f_n^{(n-1)}(x_0)
\end{aligned}
\right\} \qquad (1-60)
$$

can therefore be solved for the n constants $\tilde{c}_1$, $\tilde{c}_2$, . . ., $\tilde{c}_n$. Hence, the linear superposition principle shows that the function $g(x)$ defined by

$$g(x) = \tilde{c}_1 f_1(x) + \tilde{c}_2 f_2(x) + \ . \ . \ . + \tilde{c}_n f_n(x) \qquad (1-61)$$

is a solution to equation (1–55) on the interval (1–56). And it follows from equation (1–60) that

$$h(x_0) = g(x_0), h'(x_0) = g'(x_0), \ . \ . \ ., h^{(n-1)}(x_0) = g^{(n-1)}(x_0)$$

But since the fundamental theorem shows that there is only one solution satisfying a set of n initial conditions, we can conclude that $g(x) = h(x)$ for all x on the interval (1–56). Equation (1–61) now shows that equation (1–59) holds, and this proves the assertion.

It is easy to see from these results that equation (1–56) cannot possess $n+1$ linearly independent solutions on the interval (1–56).

Now consider the general linear equation (1–7) and suppose that the coefficients $a_0(x)$, $a_1(x)$, . . ., $a_n(x)$, $b(x)$ are continuous on the interval

$$\alpha \leqslant x \leqslant \beta \qquad (1\text{–}62)$$

and that $a_0(x)$ *does not vanish at any point of this interval. Let $f_0(x)$ be any particular solution of equation (1–7); and let $f_1, f_2, \ldots, f_n$ be any n linearly independent solutions of the associated homogeneous equation. Then if* h(x) *satisfies equation (1–7) on this interval we can find* n *constants* $\tilde{c}_1, \ldots, \tilde{c}_n$ *such that*

$$h(x) = \tilde{c}_1 f_1(x) + \ldots + \tilde{c}_n f_n(x) + f_0(x) \qquad (1\text{–}63)$$

on the entire interval.

This means that every solution on the interval can be obtained from the general solution

$$f(x, c_1, c_2, \ldots, c_n) = c_1 f_1(x) + c_2 f_2(x) + \ldots + c_n f_n(x) + f_0(x) \qquad (1\text{–}64)$$

In order to prove this assertion, notice that $h(x) - f_0(x)$ is a solution to the associated homogeneous equation of equation (1–7). Hence, we can find constants $\tilde{c}_1, \tilde{c}_2, \ldots, \tilde{c}_n$ such that

$$h(x) - f_0(x) = \tilde{c}_1 f_1(x) + \tilde{c}_2 f_2(x) + \ldots + \tilde{c}_n f_n(x)$$

This shows that equation (1–64) holds and proves the assertion.

The general solution of the associated homogeneous equation

$$c_1 f_1(x) + c_2 f_2(x) + \ldots + c_n f_n(x)$$

which appears in the general solution (1–64) of the nonhomogeneous equation (1–7) is called the *complementary function.* And any set of n linearly independent homogeneous solutions is known as a set of *fundamental solutions* to the homogeneous equation.

For example, the complementary function of the second-order equation

$$\frac{d^2y}{dx^2} + y = x \tag{1-65}$$

is $c_1 \cos x + c_2 \sin x$ where c_1 and c_2 are arbitrary constants. A particular solution to this equation is $y = x$. Then

$$y = c_1 \cos x + c_2 \sin x + x$$

is a general solution of equation (1-65), and any other particular solution of this equation can be obtained by assigning particular numerical values to the constants c_1 and c_2.

CHAPTER 2

First-Order Equations

In this chapter a number of elementary techniques for obtaining explicit solutions for certain types of first-order equations are developed. These equations are chosen for consideration because they are simple enough to be solved explicitly and, at the same time, occur frequently enough in practice so that they are worth solving. As it happens, these equations are all of the first degree.

2.1 SYMMETRIC NOTATION FOR EQUATIONS OF FIRST DEGREE

The first-order differential equation of the first degree is an equation of the form

$$y'N(x, y) + M(x, y) = 0 \qquad (2\text{--}1)$$

We shall suppose that M and N possess continuous first partial derivatives in some domain D of the x, y plane. If $N(x, y)$ were nonzero at every point of D, we could write equation (2–1) in the normal form

$$y' = f(x, y) \equiv -\frac{M(x, y)}{N(x, y)} \qquad (2\text{--}2)$$

and the fundamental theorem would guarantee that this equation have a unique solution passing through each point of D. Points of D where $N(x, y) = 0$ are called *singular points* of the equation.

In equation (2–1), y is considered as the dependent variable. The solutions of this equation are functions of the form

$$y = f(x) \tag{2-3}$$

On the other hand, the equation

$$\frac{dx}{dy} M(x, y) + N(x, y) = 0 \tag{2-4}$$

which is closely related to equation (2–1) possesses solutions which are functions of the form

$$x = g(y) \tag{2-5}$$

The singular points of this equation occur when $M(x, y) = 0$.

However, in any domain in which neither equation (2–1) nor (2–4) has a singular point, any solution (2–3) of equation (2–1) can be solved for x to obtain a solution of the form (2–5) of equation (2–4), and conversely. Thus, for many purposes, it is not necessary to distinguish between equations (2–1) and (2–2). When this is the case, we can introduce the symmetrical notation

$$M(x, y)dx + N(x, y)dy = 0 \tag{2-6}$$

If $M(x, y)$ and $N(x, y)$ are both not zero, we understand this notation to denote either equation (2–1) or equation (2–4). When $M(x, y) = 0$ but $N(x, y) \neq 0$, the existence of the solution (2–3) of equation (2–1) is guaranteed by the fundamental theorem, but that of equation (2–4) is not.[25] Hence, in the neighborhood of a point where $M(x, y) = 0$ we understand the notation (2–6) to denote equation (2–1). Similarly, when $N(x, y) = 0$ but $M(x, y) \neq 0$, we understand the notation (2–6) to denote equation (2–4). The fundamental theorem does not guarantee that either equation will possess solutions at points where $M(x, y)$ and $N(x, y)$ are both zero, and we therefore say that such points are singular points of equation (2–6).

---

[25] At such points, $dx/dy = \infty$ and $dy/dx = 0$; hence, dx/dy is not continuous.

2.2 EXACT EQUATIONS

The first-order differential equation of the first degree

$$M(x, y)\,dx + N(x, y)\,dy = 0 \qquad (2\text{-}7)$$

with M and N defined and continuous in some domain D, is said to be *exact* in D if the line integral

$$\oint [M(x, y)\,dx + N(x, y)\,dy]$$

has the same value for all paths of integration which lie in D and which have the same end points. This is equivalent to requiring that the integral vanish around every closed path within D.

It is shown in books on calculus that the differential equation (2-7) is exact if, and only if, there exists a continuously differentiable function $\phi(x, y)$ such that[26]

$$M = \frac{\partial \phi}{\partial x} \qquad \text{and} \qquad N = \frac{\partial \phi}{\partial y} \qquad (2\text{-}8)$$

Then the function ϕ, which is defined only to within an additive constant, is given by

$$\phi(x, y) = \oint_{(x_0, y_0)}^{(x, y)} (M\,dx + N\,dy) + \text{constant} \qquad (2\text{-}9)$$

where (x_0, y_0) is any fixed point and the integral is carried out over any curve Γ within D which joins the fixed point (x_0, y_0) with the variable point (x, y).

Suppose that equation (2-7) is exact. Then it follows from equation (2-8) that the total derivative of ϕ with respect to x along any direction is

$$\frac{d\phi}{dx} = \frac{\partial \phi}{\partial x} + \frac{\partial \phi}{\partial y}\,y' = M + y'N \qquad (2\text{-}10)$$

Similarly, the total derivative with respect to y is

---

[26] Alternatively, we can say that $M\,dx + N\,dy$ is an exact differential of the function ϕ or that $d\phi = M\,dx + N\,dy$.

$$\frac{d\phi}{dy} = \frac{\partial\phi}{\partial y} + \frac{\partial\phi}{\partial x}\frac{dx}{dy} = N + \frac{dx}{dy}M \qquad (2\text{--}11)$$

Since the differential equation (2–7) denotes either the equation

$$M + y'N = 0$$

or the equation

$$N + \frac{dx}{dy}M = 0$$

equations (2–10) and (2–11) show that ϕ is constant along every solution curve of equation (2–7). Hence, ϕ is an integral of equation (2–7). However, for a first-order equation, finding its integral is equivalent to finding its solution in implicit form. Thus, an exact equation can effectively be solved simply by carrying out the integral (2–9) along any convenient path.

It is easy to see by differentiating equations (2–8) and interchanging the order of differentiation that, if M and N are continuously differentiable, a necessary condition for equation (2–7) to be exact is that

$$\frac{\partial M}{\partial y} = \frac{\partial N}{\partial x} \qquad (2\text{--}12)$$

Now suppose that the domain D has no holes. That is, it is similar to the domain shown in figure 2–1(a) but not the one shown in figure 2–1(b). Such a domain is said to be *simply connected*.

Let Γ be any closed curve within D and let R be the region enclosed by Γ, as shown schematically in figure 2–2. Then Green's theorem

$$\int_{\Gamma} [M\,dx + N\,dy] = \iint_{R} \left(\frac{\partial N}{\partial x} - \frac{\partial M}{\partial y}\right) dx\,dy$$

shows that condition (2–12) is also a sufficient condition for exactness. If D were not simply connected, condition (2–12) would not be sufficient to ensure that the line integral will always vanish along a closed path such as Γ in figure 2–2.

 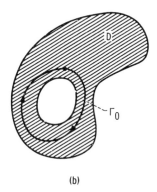

(a) Domain has no holes.   (b) Domain has a single hole.

FIGURE 2–1.—Singly and multiply connected domains.

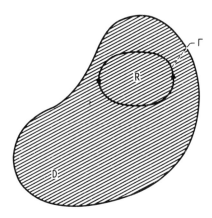

FIGURE 2–2.—Path of integration within D.

We shall always suppose that the conditions of differentiability and connectedness given in the preceding paragraphs are satisfied and therefore that equation (2–12) is the necessary and sufficient condition for equation (2–7) to be exact.

For example, consider the equation

$$\frac{2x-y}{x^2+y^2}\,dx + \frac{2y+x}{x^2+y^2}\,dy = 0$$

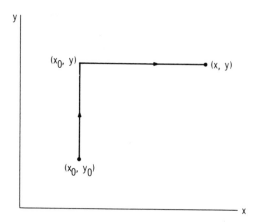

FIGURE 2–3. – Path of integration for finding integral.

Then

$$\frac{\partial M}{\partial y} \equiv \frac{\partial}{\partial y}\left(\frac{2x-y}{x^2+y^2}\right) = \frac{y^2-x^2-4xy}{(x^2+y^2)^2}$$

and

$$\frac{\partial N}{\partial x} \equiv \frac{\partial}{\partial x}\left(\frac{2y+x}{x^2+y^2}\right) = \frac{y^2-x^2-4xy}{(x^2+y^2)^2}$$

Hence, equation (2–12) holds and the differential equation is exact. Therefore, carrying out the integral (2–9) along the path depicted in figure 2–3 shows that

$$\phi(x, y) = \int_{y_0}^{y} \frac{2y+x_0}{x_0^2+y^2}\,dy + \int_{x_0}^{x} \frac{2x-y}{x^2+y^2}\,dx$$

$$= \ln(x^2+y^2) - \tan^{-1}\frac{x}{y} + \text{constant}$$

is an integral of the equation.

Unfortunately, the majority of first-degree equations encountered in practice are not exact. It is therefore natural to ask whether there exists a function $\lambda(x, y)$ such that the equation

$$\lambda(M\,dx + N\,dy) = 0 \tag{2–13}$$

(obtained by multiplying both sides of eq. (2–7) by λ) is exact. It can be shown (ref. 3, p. 27) that such a function always exists, provided only that equation (2–7) possesses exactly one general solution. It is called an *integrating factor*. Integrating factors are certainly not unique for, if λ is an integrating factor of a given first-order equation, so is cλ for any constant c.

Now if λ is an integrating factor for equation (2–7), the condition (2–12) for exactness must hold for equation (2–13). Hence,

$$\frac{\partial(\lambda M)}{\partial y} = \frac{\partial(\lambda N)}{\partial x}$$

or

$$M\frac{\partial\lambda}{\partial y} - N\frac{\partial\lambda}{\partial x} + \lambda\left(\frac{\partial M}{\partial y} - \frac{\partial N}{\partial x}\right) = 0 \qquad (2-14)$$

Thus, λ is the solution of a linear partial differential equation (see section 3.3) which is usually more difficult to solve than the original ordinary differential equation. However, since it can be shown that every solution of equation (2–14) is an integrating factor of equation (2–7), it is only necessary to find a single particular solution to equation (2–14).

Although, in general, finding an integrating factor is quite difficult, it is sometimes possible to accomplish this simply by inspection. For example, upon multiplying the equation

$$[f(x) + y]dx - x\,dy = 0$$

by $1/x^2$ we obtain the equation

$$\frac{1}{x^2}[f(x) + y]dx - \frac{1}{x}dy = 0$$

And since $\partial/\partial y[(f+y)/x^2] = 1/x^2$ and $\partial/\partial x(-1/x) = 1/x^2$, this equation is exact. Then carrying out the line integral (2–9) shows that the differential equation possesses the integral

$$\phi = -\frac{y}{x} + \int \frac{f(x)}{x^2}\,dx$$

2.3 EQUATIONS OF THE TYPE $g(y)y'=f(x)+h(x)G\left(\int f(x)dx-\int g(y)dy\right)$

Many of the differential equations which can be solved by classical methods are special cases of the differential equation

$$-g(y)dy+\left[f(x)+h(x)G\left(\int f(x)dx-\int g(y)dy\right)\right]dx=0 \qquad (2\text{-}15)$$

where $g(y), f(x), h(x)$, and G can be any functions of their arguments. In order to obtain a solution to this equation put

$$U\equiv\int f(x)dx-\int g(y)dy$$

Then $dU=f(x)dx-g(y)dy$ and equation (2–15) can be written as

$$dU+h(x)G(U)dx=0$$

It is easy to see that $1/G(U)$ is an integrating factor for this differential equation and therefore that the equation

$$\frac{dU}{G(U)}+h(x)dx=0$$

is exact. Hence, it follows from equation (2–9) that

$$\phi\equiv\int^{\int f(x)dx-\int g(y)dy}\frac{1}{G(U)}dU+\int h(x)dx=\text{constant} \qquad (2\text{-}16)$$

is an integral of equation (2–15).

We now show that many of the first-order normal equations, which are solvable by the classical methods, can be obtained by specializing the functions f, g, h, and G in equation (2–15) and, therefore, that the solutions of these equations (actually the integrals) are all given by equation (2–16).

First put $G(U)=1$, and $h(x)=0$. Then equation (2-15) becomes the general *separable equation*

$$-g(y)dy+f(x)dx=0$$

and equation (2-16) shows that its integral is simply

$$\phi=\int f(x)dx-\int g(y)dy$$

Now put $G(U) \equiv (\alpha/\beta)e^{U/\alpha}H(e^{-U})-1$, $g(y) \equiv \alpha/y$, $f(x)=h(x) \equiv \beta/x$. Then equation (2-15) becomes

$$dy= x^{(\beta/\alpha)-1}H\left(\frac{y^\alpha}{x^\beta}\right)dx \tag{2-17}$$

which is the *isobaric* (or *one-dimensional*) equation. In this case, the integral (2-16) becomes

$$\int^{\ln(x^\beta/y^\alpha)} \frac{dU}{\dfrac{\alpha}{\beta}e^{U/\alpha}H(e^{-U})-1}-\ln x^\beta=\text{constant}$$

or

$$\int^{(y^\alpha/x^\beta)} \frac{dV}{\dfrac{\alpha}{\beta}V^{1-(1/\alpha)}H(V)-V}-\ln x^\beta=\text{constant}$$

If we put $\alpha=\beta=1$ in equation (2-17), we obtain the *homogeneous equation*

$$dy=H\left(\frac{y}{x}\right)dx \tag{2-18}$$

And equation (2-16) shows that its solution is

$$\int^{y/x}\frac{dV}{H(V)-V}-\ln x=\text{constant}$$

In order to recognize whether a given differential equation has the form (2–17), it is necessary to be able to determine whether any given function $\eta(x, y)$ can be expressed in the form

$$x^{(\beta/\alpha)-1} H\left(\frac{y^\alpha}{x^\beta}\right) \tag{2–19}$$

A necessary and sufficient condition that $\eta(x, y)$ can be expressed in this form is that there exist a number p such that

$$\eta(tx, t^p y) = t^{p-1} \eta(x, y) \tag{2–20}$$

for all values of t.

In order to show that a function which satisfies the condition (2–20) can always be expressed in the form (2–19), put $t = 1/x$ in equation (2–20) to get

$$\eta\left(1, \frac{y}{x^p}\right) = x^{1-p} \eta(x, y)$$

Now choose a number α and define the number β and the function $H(U)$ by $\beta \equiv p\alpha$ and $H(U) \equiv \eta(1, U^{1/\alpha})$, respectively. Then

$$H\left(\frac{y^\alpha}{x^\beta}\right) = \eta\left[1, \left(\frac{y^\alpha}{x^\beta}\right)^{1/\alpha}\right] = \eta\left(1, \frac{y}{x^{\beta/\alpha}}\right) = \eta\left(1, \frac{y}{x^p}\right) = x^{1-(\beta/\alpha)} \eta(x, y)$$

which shows that $\eta(x, y)$ can be expressed in the form (2–19).

For example, replace y by $yt^{1/a}$ and x by xt in the function

$$\eta(x, y) = \frac{y(x - y^a a \ln y + y^a \ln x)}{x^2(a \ln y - \ln x)}$$

to show that

$$\eta(xt, yt^{1/a}) = t^{(1-a)/a} \eta(x, y)$$

Hence, this function satisfies condition (2–20), and it can be put in the form (2–19) by introducing the variable y^a/x to obtain

$$\eta(x, y) = x^{(1/a)-1} \frac{\left(\frac{y^a}{x}\right)^{1/a}\left[1 - \frac{y^a}{x}\ln\left(\frac{y^a}{x}\right)\right]}{\ln\frac{y^a}{x}}$$

In the important special case where $\alpha = \beta$ (corresponding to the homogeneous eq. (2–18)), condition (2–20) reduces to

$$\eta(tx, ty) = \eta(x, y) \tag{2–21}$$

A function of two variables which satisfies condition (2–21) is said to be homogeneous of degree zero, which, of course, accounts for the name given to equation (2–18). More generally, a function of the n variables $x_1, x_2, \ldots, x_n$, say $\eta(x_1, x_2, \ldots, x_n)$, is said to be homogeneous of degree k in the variables $x_1, x_2, \ldots, x_n$ if it satisfies the condition

$$\eta(tx_1, tx_2, \ldots, tx_n) = t^k\eta(x_1, x_2, \ldots, x_n) \tag{2–22}$$

for all values of t.

Another well-known special case of (2–15) is obtained by putting

$$G(U) = e^{-(k-1)U}, \; h(x) = \psi(x)\exp\left[(k-1)\int f(x)dx\right], \text{ and } g(y) = 1/y$$

to obtain *Bernoulli's* equation

$$dy = [f(x)y + \psi(x)y^k]dx \tag{2–23}$$

It follows from equation (2–16) that its solution is

$$\frac{\exp\left[(k-1)\int f(x)dx\right]}{(k-1)y^{k-1}} + \int \psi(x)\exp\left[(k-1)\int f(x)dx\right]dx = \text{constant} \tag{2–24}$$

When $k = 0$, equation (2–23) reduces to the first-order *linear equation*

$$dy = [f(x)y + \psi(x)]dx \tag{2–25}$$

And equation (2–24) shows that its solution is

$$y = e^{\int f(x)dx} \int e^{-\int f(x)dx} \ \psi(x)dx + \text{constant } e^{\int f(x)dx}$$

2.4 RICCATI EQUATION

The Riccati equation

$$\frac{dy}{dx} = f(x) + g(x)y + h(x)y^2 \tag{2–26}$$

can be thought of as a generalization of the linear first-order equation (2–25). This equation is sometimes called the generalized Riccati equation since Riccati actually studied the special case

$$\frac{dy}{dx} + by^2 = cx^m$$

where b and c are constants.

If a particular solution y_1 of equation (2–26) is known, the general solution y of this equation is

$$y = y_1 + \frac{1}{U}$$

where U is the general solution of the linear equation

$$\frac{dU}{dx} + (g + 2y_1h)U + h = 0$$

which was solved in section 2.3. This follows from the relation

$$\frac{dy}{dx} - f - gy - hy^2 = \frac{dy_1}{dx} - f - gy_1 - hy_1^2 - \frac{1}{U^2}\left(\frac{dU}{dx} + gU + h + 2y_1hU\right)$$

$$= -\frac{1}{U^2}\left[\frac{dU}{dx} + (g + 2y_1h)U + h\right]$$

$$= 0$$

Even if a particular solution is not known, it is still possible to reduce the problem of solving equation (2–26) to the problem of solving a linear equation; but in this case the equation is of second order. In order to accomplish this put

$$U \equiv e^{-\int yh(x)dx}$$

Then $U'/U = -yh$ and

$$U'' = -yhU' - y'hU - yh'U = (y^2h^2 - y'h)U - yh'U$$

$$= -(f + gy)hU - yh'U$$

$$= -fhU - y(gh + h')U$$

$$= -fhU + \left(g + \frac{h'}{h}\right)U'$$

Hence, U satisfies the second-order linear homogeneous differential equation

$$U'' - \left(g + \frac{h'}{h}\right)U' + fhU = 0$$

The converse of this statement is also true. That is, every second-order linear homogeneous equation can be transformed into a Riccati equation.

Since linear second-order equations will be discussed extensively in chapter 6, we shall not consider the Riccati equation further here.

CHAPTER 3

Systems of Equations

In this chapter, a number of important properties of systems of differential equations are discussed. We shall also show how the solutions of these systems can be used to obtain solutions to certain types of first-order partial differential equations.

3.1. FIRST-ORDER NORMAL SYSTEMS

In certain applications it is necessary to deal with systems of simultaneous differential equations. The set of n first-order normal differential equations in the n dependent variables $y_1, y_2, \ldots, y_n$

$$
\left.
\begin{aligned}
\frac{dy_1}{dx} &= G_1(x, y_1, y_2, \ldots, y_n) \\[1em]
\frac{dy_2}{dx} &= G_2(x, y_1, y_2, \ldots, y_n) \\[1em]
&\ \ \cdot \qquad\qquad\quad \cdot \\[0.3em]
&\ \ \cdot \qquad\qquad\quad \cdot \\[0.3em]
&\ \ \cdot \qquad\qquad\quad \cdot \\[0.3em]
\frac{dy_n}{dx} &= G_n(x, y_1, y_2, \ldots, y_n)
\end{aligned}
\right\} \tag{3-1}
$$

is called a *normal system*. It provides a standard form to which all normal differential equations and all systems of normal differential equations can be reduced. For example, by introducing the new dependent variables y_i defined by

$$y_{i+1} \equiv \frac{d^i y}{dx^i} \qquad \text{for } i = 1, 2, \ldots n-1$$

the nth-order normal equation $(1-19)$ can be transformed into a first-order normal system

$$\frac{dy_1}{dx} = y_2$$

$$\frac{dy_2}{dx} = y_3$$

$$\cdot \qquad \cdot$$
$$\cdot \qquad \cdot$$
$$\cdot \qquad \cdot$$

$$\frac{dy_{n-1}}{dx} = y_n$$

$$\frac{dy_n}{dx} = G(x, y_1, y_2, \ldots, y_n)$$

However, it is by no means always possible to transform a first-order normal system into a single nth-order equation.

In order to obtain a geometric interpretation of the system $(3-1)$ and its solutions, it is again convenient to think of the variables $x, y_1, y_2, \ldots, y_n$ as being the coordinates of an $(n+1)$-dimensional space. The definitions of a domain and of a neighborhood in this space were given in section 1.1. Based on these ideas, the fundamental theorem for the first-order normal system can be stated as follows:

Suppose that the functions G_i *for* $i = 1, 2, \ldots, n$ *are defined and continuous in some domain* D *and that the partial derivatives*

$$\frac{\partial G_i}{\partial y_j} \qquad \text{for } 1 \leq i \leq n;\ 1 \leq j \leq n$$

are continuous in D. *Then for each point* $(x_0, y_1^0, y_2^0, \ldots, y_n^0)$ *of* D *equation* *(3-1) has precisely one solution*

$$\left.\begin{aligned}
y_1 &= f_1(x) \\
y_2 &= f_2(x) \\
&\quad . \qquad . \\
&\quad . \qquad . \\
&\quad . \qquad . \\
y_n &= f_n(x)
\end{aligned}\right\} \tag{3-2}$$

which is defined on some interval $\alpha < \mathrm{x} < \beta$ containing the point x_0 and satisfies the initial conditions

$$f_1(x_0) = y_1^0, \; f_2(x_0) = y_2^0, \; \ldots, \; f_n(x_0) = y_n^0 \tag{3-3}$$

The solutions of the system (3-1) can be visualized as curves in the $(n+1)$-dimensional space. Then the fundamental theorem states that there is precisely one such curve passing through each point of the domain D in this space.

If $H(x, y_1, y_2, \ldots, y_n)$ is a nonconstant function such that every solution to the system (3-1) satisfies an equation of the form[27]

$$H(x, y_1, y_2, \ldots, y_n) = \text{constant} \tag{3-4}$$

(the constant may be different for different solutions), then $H(x, y_1, y_2 \ldots, y_n)$ is called an *integral* (or *first integral*) of the system (3-1). Thus, an integral of the system (3-1) is a function H which is not identically constant but is constant along each solution curve of the system. Therefore, upon differentiating H along any solution curve of the system (3-1) we obtain

$$\frac{dH}{dx} = \frac{\partial H}{\partial x} + \frac{\partial H}{\partial y_1}\frac{dy_1}{dx} + \ldots + \frac{\partial H}{\partial y_n}\frac{dy_n}{dx}$$

$$= \frac{\partial H}{\partial x} + \frac{\partial H}{\partial y_1} G_1 + \ldots + \frac{\partial H}{\partial y_n} G_n = 0$$

---

[27] Notice that this definition of an integral is consistent with the one given in chapter 1 for the nth-order normal equation when the latter equation is transformed, by the procedure just described, into a first-order normal system.

provided that H possesses continuous first partial derivatives. But each point of D lies on a solution curve of the system (3–1). Hence, H satisfies the equation

$$\frac{\partial H}{\partial x} + \frac{\partial H}{\partial y_1} G_1 + \frac{\partial H}{\partial y_2} G_2 + \ldots + \frac{\partial H}{\partial y_n} G_n = 0 \tag{3–5}$$

at every point of D. Conversely, if H is any solution to equation (3–5) in D, it is certainly constant along every solution curve of the system (3–1). Hence, H must be an integral of this system. We have now shown that a nonconstant function H with continuous partial derivatives is an integral of the system (3–1) in D if, and only if, it satisfies the first-order linear partial differential equation (3–5).

Each integral H of the system (3–1) determines a family of n-dimensional hypersurfaces in the region D, and every solution curve lies on one of these hypersurfaces. Now suppose that we have found n integrals, say H_1, H_2, $\ldots$, H_n, of the system (3–1) which are defined in the region D. If the *Jacobian* determinant

$$\frac{\partial (H_1, \ldots, H_n)}{\partial (y_1, \ldots, y_n)} \equiv \begin{vmatrix} \dfrac{\partial H_1}{\partial y_1} & \cdots & \dfrac{\partial H_1}{\partial y_n} \\ \cdot & & \cdot \\ \cdot & & \cdot \\ \cdot & & \cdot \\ \dfrac{\partial H_n}{\partial y_1} & \cdots & \dfrac{\partial H_n}{\partial y_n} \end{vmatrix} \tag{3–6}$$

of these integrals is different from zero in D, it will be possible[28] to solve the n equations

$$H_1(x, y_1, \ldots, y_n) = c_1, \ldots, H_n(x, y_1, \ldots, y_n) = c_n \tag{3–7}$$

for $y_1, \ldots, y_n$ as functions of x. And these functions will depend on the n arbitrary constants $c_1, \ldots, c_n$. The n equations (3–7) therefore implicitly determine an n-parameter family of curves in the $(n+1)$-dimensional space.

---

[28] This is a consequence of the implicit function theorem for systems of equations. See, for example, ref. 6.

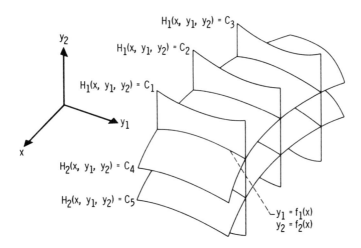

FIGURE 3–1. — Relation between integral curves and surfaces of ordinary differential equations.

But these curves must lie along the various mutual intersections of n hypersurfaces

$$H_1 = \text{constant}, \qquad H_2 = \text{constant}, \qquad \ldots, \qquad H_n = \text{constant}$$

This is shown schematically in figure 3–1 for the case where $n = 2$. However, since every solution curve of the system (3–1) must simultaneously lie on n hypersurfaces, each of which corresponds to one of the n integrals $H_1, \ldots, H_n$, the solution curves must also lie along these intersections. Therefore, every solution to the system (3–1) can be obtained as an implicit solution of the n equations (3–7). That is, the n integrals $H_1, \ldots, H_n$ which have a non-vanishing Jacobian in D are just sufficient to completely determine all the solutions of the systems (3–1) in D. However, the requirement that the Jacobian (3–6) be different from zero in D implies that the integrals are functionally independent [29] in D. We therefore anticipate that the system (3–1) will possess n functionally independent integrals. This turns out to be the case, at least when the region D is chosen to be sufficiently small (see ref. 7).

---

[29] The functions $H_1, \ldots, H_n$ are functionally dependent in D if there exists a nonconstant function ω of $H_1, \ldots, H_n$ such that $\omega(H_1, \ldots, H_n) = 0$ for all $x, y_1, \ldots, y_n$ in D. See ref. 6.

For example, consider the normal system

$$\frac{dy_1}{dx} = G_1(x, y_1, y_2) \equiv \frac{xy_1}{y_1^2 - y_2^2}$$

$$\frac{dy_2}{dx} = G_2(x, y_1, y_2) \equiv \frac{xy_2}{y_2^2 - y_1^2}$$

which is defined in some region D which does not include the planes $y_1 = \pm y_2$. Since the two functions $H_1(x, y_1, y_2) = y_1 y_2$ and $H_2 = y_1^2 + y_2^2 - x^2$ satisfy the equations

$$\frac{\partial H_1}{dx} + \frac{\partial H_1}{\partial y_1} G_1 + \frac{\partial H_1}{\partial y_2} G_2 = 0 + y_2 \left(\frac{xy_1}{y_1^2 - y_2^2}\right) + y_1 \left(\frac{xy_2}{y_2^2 - y_1^2}\right) \equiv 0$$

and

$$\frac{\partial H_2}{\partial x} + \frac{\partial H_2}{\partial y_1} G_1 + \frac{\partial H_2}{\partial y_2} G_2 = -2x + 2y_1 \left(\frac{xy_1}{y_1^2 - y_2^2}\right) + 2y_2 \left(\frac{xy_2}{y_2^2 - y_1^2}\right) \equiv 0$$

we know that H_1 and H_2 are two integrals of this system. Since

$$\frac{\partial(H_1, H_2)}{\partial(y_1, y_2)} = \begin{vmatrix} y_2 & y_1 \\ 2y_1 & 2y_2 \end{vmatrix} = 2(y_2^2 - y_1^2)$$

the Jacobian of these functions does not vanish in D. And this shows that H_1 and H_2 are functionally independent. The solution curves lie on the intersections of the surfaces $H_1 = $ constant with the surfaces $H_2 = $ constant. Thus, every solution curve lies on an intersection of a hyperbolic cylinder and a hyperboloid (or cone).

Now let $H_1, H_2, \ldots, H_{n+1}$ be any $n+1$ integrals of the system (3–1) in some domain D. Since each of these integrals must satisfy equation (3–5), it follows that

$$\frac{\partial H_1}{\partial x} + \frac{\partial H_1}{\partial y_1} G_1 + \ldots + \frac{\partial H_1}{\partial y_n} G_n = 0$$

$$\frac{\partial H_2}{\partial x} + \frac{\partial H_2}{\partial y_1} G_1 + \ldots + \frac{\partial H_2}{\partial y_n} G_n = 0$$

$$\frac{\partial H_{n+1}}{\partial x} + \frac{\partial H_{n+1}}{\partial y_1} G_1 + \ldots + \frac{\partial H_{n+1}}{\partial y_n} G_n = 0$$

And it must be possible to solve these equations for $1, G_1, G_2, \ldots, G_n$ at every point of D. However, this can happen only if the determinant of the coefficients

$$\frac{\partial(H_1, H_2, \ldots, H_{n+1})}{\partial(x, y_1, \ldots, y_n)} = \begin{vmatrix} \dfrac{\partial H_1}{\partial x} & \dfrac{\partial H_1}{\partial y_1} & \cdots & \dfrac{\partial H_1}{\partial y_n} \\ \dfrac{\partial H_2}{\partial x} & \dfrac{\partial H_2}{\partial y_1} & \cdots & \dfrac{\partial H_2}{\partial y_n} \\ \cdot & \cdot & & \cdot \\ \cdot & \cdot & & \cdot \\ \cdot & \cdot & & \cdot \\ \dfrac{\partial H_{n+1}}{\partial x} & \dfrac{\partial H_{n+1}}{\partial y_1} & \cdots & \dfrac{\partial H_{n+1}}{\partial y_n} \end{vmatrix}$$

is zero at every point of D. But this means that $H_1, \ldots, H_{n+1}$ are functionally dependent. Hence, every set of $n+1$ integrals of the system (3–1) is functionally dependent. We have, therefore, shown that *the system (3–1) possesses precisely* n *functionally independent integrals* (*at least in a sufficiently small region* D).

If n functions $I_1(x, y_1, \ldots, y_n), \ldots, I_n(x, y_1, \ldots, y_n)$ not all identically zero can be found such that

$$I_1\left(\frac{dy_1}{dx} - G_1\right) + \ldots + I_n\left(\frac{dy_n}{dx} - G_n\right)$$

is equal to dF/dx for some nonconstant function $F(x, y_1, \ldots, y_n)$, then F will be constant along every solution curve of the system (3–1). Hence, F will be an integral of the system (3–1). The functions $I_1, \ldots, I_n$ can be thought of as integrating factors.[30] Although it is usually difficult to find integrating factors, they can sometimes be guessed at from the symmetry of the problem.

For example, consider the system

$$
\left.
\begin{aligned}
\frac{dy_1}{dx} &= y_2 - y_3 \\[2mm]
\frac{dy_2}{dx} &= y_3 - y_1 \\[2mm]
\frac{dy_3}{dx} &= y_1 - y_2
\end{aligned}
\right\}
\qquad (3\text{–}8)
$$

Then choosing $I_1 = I_2 = I_3 = 1$ gives

$$
1 \times \left[\frac{dy_1}{dx} - (y_2 - y_3) \right] + 1 \times \left[\frac{dy_2}{dx} - (y_3 - y_1) \right] + 1 \times \left[\frac{dy_3}{dx} - (y_1 - y_2) \right]
$$

$$
= \frac{dy_1}{dx} + \frac{dy_2}{dx} + \frac{dy_3}{dx} = \frac{d}{dx}(y_1 + y_2 + y_3)
$$

Hence, $H_1 = y_1 + y_2 + y_3$ is an integral of the system (3–8). Now choose $I_1 = 2y_1$, $I_2 = 2y_2$, and $I_3 = 2y_3$. Then

$$
2y_1 \left[\frac{dy_1}{dx} - (y_2 - y_3) \right] + 2y_2 \left[\frac{dy_2}{dx} - (y_3 - y_1) \right] + 2y_3 \left[\frac{dy_3}{dx} - (y_1 - y_2) \right]
$$

$$
= 2y_1 \frac{dy_1}{dx} + 2y_2 \frac{dy_2}{dx} + 2y_3 \frac{dy_3}{dx} = \frac{d}{dx}(y_1^2 + y_2^2 + y_3^2)
$$

Hence, $H_2 = y_1^2 + y_2^2 + y_3^2$ is also an integral of the system (3–8). It is easy to verify that H_1 and H_2 are functionally independent.

---

[30] See section 2.2.

Instead of interpreting the solution (3–2) of the normal system (3–1) as the explicit equation of a curve in an $(n+1)$-dimensional space with coordinates $x, y_1, \ldots, y_n$, we can interpret it as the parametric equation (with parametric variable x) of a curve in an n-dimensional space with coordinates $y_1, \ldots, y_n$. The parametric variable x may be the arc length or a similar parameter for the solution curves (such as the time in a physical problem). The fundamental theorem can still be used to show that there is a solution curve passing through every point of any region of this n-dimensional space, provided the functions G_i satisfy the appropriate smoothness requirements. However, except in a certain important special case (which is the topic of the next section), it will usually not be true that only a single curve passes through each point of this region.

The simplicity of the first-order normal systems becomes apparent when equation (3–1) is written in vector notation. To this end we notice that the coordinates of a point $y_1, \ldots, y_n$ in the n-dimensional space can be interpreted as the components of an n-dimensional vector $\mathbf{y}$ and we write [31]

$$\mathbf{y} = (y_1, y_2, \ldots, y_n)$$

A vector is said to be a function of a single real variable x if its components are functions of x. Thus, if (as in the case of the solution curve (3–2) of the system (3–1)),

$$y_1 = f_1(x), \ldots, y_n = f_n(x)$$

we write

$$\mathbf{y} = \mathbf{f}(x) \equiv (f_1(x), \ldots, f_n(x))$$

---

[31] Addition of two vectors and multiplication of a vector by a scalar are defined component-wise. Thus, if $\mathbf{y} = (y_1, \ldots, y_n)$ and $\mathbf{x} = (x_1, \ldots, x_n)$ are vectors and a is a number, $\mathbf{y} + \mathbf{x} = ((y_1 + x_1), \ldots, (y_n + x_n))$, $a\mathbf{y} = (ay_1, \ldots, ay_n)$. The length or magnitude of the vector $\mathbf{y}$ is denoted by $|\mathbf{y}|$ and is defined as the distance from the origin to the point $y_1, \ldots, y_n$. Thus, $|\mathbf{y}| = (y_1^2 + \ldots + y_n^2)^{1/2}$. Notice that the magnitude of a vector is zero if, and only if, its components are all zeros. The *dot product* or *inner product* of two vectors $\mathbf{y}$ and $\mathbf{x}$ is denoted by $\mathbf{y} \cdot \mathbf{x}$ and is defined by $\mathbf{y} \cdot \mathbf{x} = y_1 x_1 + \ldots + y_n x_n$. The dot product satisfies the Schwarz inequality $|\mathbf{y} \cdot \mathbf{x}| \leq |\mathbf{y}| \, |\mathbf{x}|$, where the vertical lines on the left are the absolute value of a pure number and those on the right denote the magnitude of vectors. Thus, if $|\mathbf{y}| \neq 0$ and $|\mathbf{x}| \neq 0$ then $|\mathbf{y} \cdot \mathbf{x}|/|\mathbf{y}| \, |\mathbf{x}| \leq 1$ and this quantity can be interpreted as the cosine of the angle between the vectors $\mathbf{y}$ and $\mathbf{x}$. If $\mathbf{y} \cdot \mathbf{x} = 0$, we say that $\mathbf{y}$ and $\mathbf{x}$ are orthogonal; and if $|\mathbf{y} \cdot \mathbf{x}|/|\mathbf{y}| \, |\mathbf{x}| = 1$, we say that the vectors are in the same direction. The vector $(1, 0, 0, \ldots, 0)$ has unit magnitude and is in the direction of the y_1-coordinate axis. The vector $(0, 1, 0, 0, \ldots, 0)$ has unit magnitude and is in the direction of the y_2-coordinate axis, etc. We denote these vectors by $\hat{\mathbf{k}}_1, \hat{\mathbf{k}}_2$, etc. For more information, see ref. 8.

DIFFERENTIAL EQUATIONS

Differentiation of vectors is defined component-wise, and we write

$$\frac{d\mathbf{y}}{dx} \equiv \left(\frac{dy_1}{dx}, \ldots, \frac{dy_n}{dx}\right) = \frac{d\mathbf{f}(x)}{dx} \equiv \left(\frac{df_1(x)}{dx}, \ldots, \frac{df_n(x)}{dx}\right)$$

If $f(y_1, \ldots, y_n)$ is a function of the n variables $y_1, \ldots, y_n$ and $\mathbf{y} = (y_1, \ldots, y_n)$ denotes the vector corresponding to these variables, we write

$$f(y_1, \ldots, y_n) = f(\mathbf{y})$$

and say that f is a function of the vector variable $\mathbf{y}$. More generally, we say that a vector is a function of a vector variable if its components are. Thus, if $y_1 = f_1(\mathbf{x}), \ldots, y_n = f_n(\mathbf{x})$, then

$$\mathbf{y} = \mathbf{f}(\mathbf{x}) = (f_1(\mathbf{x}), \ldots, f_n(\mathbf{x}))$$

is a vector function of the vector variable $\mathbf{x}$.

Notice that there is no such thing as the derivative of a function of a vector variable with respect to that variable, but only partial derivatives with respect to the components of the vector variable. Thus, if $\hat{\mathbf{k}}_1, \ldots, \hat{\mathbf{k}}_n$ denote the unit vectors in the directions of the coordinate axes, we define the vector gradient operator ∇ by

$$\nabla f(\mathbf{x}) = \hat{\mathbf{k}}_1 \frac{\partial f(\mathbf{x})}{\partial x_1} + \ldots + \hat{\mathbf{k}}_n \frac{\partial f(\mathbf{x})}{\partial x_n}$$

Just as in the three-dimensional case, the effect of operating with the gradient operator on a scalar produces a vector.

A vector function is said to be continuous if its components are continuous. Thus, in the system (3–1), each G_i is a function of a real variable x and a vector variable $\mathbf{y}$ and we write

$$G_i(x, y_1, \ldots, y_n) = G_i(x, \mathbf{y})$$

If we let $\mathbf{G}$ be the vector whose components are G_i, then $\mathbf{G}$ is a vector function of the real variable x and the vector variable $\mathbf{y}$ and

$$\mathbf{G}(x, \mathbf{y}) = (G_1(x, \mathbf{y}), \ . \ . \ ., G_n(x, \mathbf{y}))$$

Hence, upon using the component-wise definition of differentiation of vectors, the first-order normal system (3–1) can be written in the concise form

$$\frac{d\mathbf{y}}{dx} = \mathbf{G}(x, \mathbf{y})$$

When the first-order system (3–1) is written in this form, the analogy between this system and a single first-order equation becomes particularly apparent.

3.2 AUTONOMOUS SYSTEMS

When the functions G_i in the system (3–1) do not involve the variable x explicitly, the system is said to be *autonomous* and it can be written in the form

$$
\left.
\begin{aligned}
\frac{dy_1}{dx} &= G_1(y_1, y_2, \ . \ . \ ., y_n) \\
\frac{dy_2}{dx} &= G_2(y_1, y_2, \ . \ . \ ., y_n) \\
&\ \ \vdots \\
\frac{dy_n}{dx} &= G_n(y_1, y_2, \ . \ . \ ., y_n)
\end{aligned}
\right\}
\qquad (3\text{–}9)
$$

If the independent variable x is thought of as representing time, autonomous systems can be interpreted as time-independent, or stationary, systems.

The first-order normal system of $n - 1$ equation

$$
\left.
\begin{aligned}
\frac{dy_1}{dx} &= G_1(y_1, y_2, \ . \ . \ ., y_{n-1}, x) \\
\frac{dy_2}{dx} &= G_2(y_1, y_2, \ . \ . \ ., y_{n-1}, x) \\
&\ \ \vdots \\
\frac{dy_{n-1}}{dx} &= G_{n-1}(y_1, y_2, \ . \ . \ ., y_{n-1}, x)
\end{aligned}
\right\}
\qquad (3\text{–}10)
$$

can always be transformed into an autonomous system in n variables which has the property that at least one of the functions on the right side never vanishes. In order to do this, it is only necessary to introduce a new dependent variable y_n by

$$y_n = x$$

and then rewrite the system (3–10) in the form

$$\frac{dy_1}{dx} = G_1(y_1, y_2, \ldots, y_n)$$

$$\frac{dy_2}{dx} = G_2(y_1, y_2, \ldots, y_n)$$

$$\vdots \qquad \vdots$$

$$\frac{dy_{n-1}}{dx} = G_{n-1}(y_1, y_2, \ldots, y_n)$$

$$\frac{dy_n}{dx} = 1$$

Conversely, if one of the G_i's of the autonomous system (3–9) in n variables does not vanish in some domain D_0, it can be transformed into a nonautonomous system in $n-1$ variables plus an additional equation which can be solved to determine the variable x once the remaining variables have been found. We may assume without loss of generality that the notation has been so chosen that G_n does not vanish in D_0. Now in order to accomplish this transformation put

$$\tilde{x} \equiv y_n \qquad\qquad (3\text{--}11)$$

and

$$\tilde{G}_i = \frac{G_i}{G_n} \qquad \text{for } 1 \leqslant i \leqslant (n-1) \qquad\qquad (3\text{--}12)$$

Then the system (3–9) becomes

$$\frac{dy_1}{d\tilde{x}}=\tilde{G}_1(y_1, \ldots, y_{n-1}, \tilde{x})$$

$$\frac{dy_2}{d\tilde{x}}=\tilde{G}_2(y_1, \ldots, y_{n-1}, \tilde{x})$$

$$\cdot \qquad \qquad \cdot$$

$$\cdot \qquad \qquad \cdot$$

$$\cdot \qquad \qquad \cdot$$

$$\frac{dy_{n-1}}{d\tilde{x}}=\tilde{G}_{n-1}(y_1, \ldots, y_{n-1}, \tilde{x}) \qquad \qquad (3\text{--}13)$$

and the equation which determines x once the system (3–13) is solved is

$$\frac{dx}{d\tilde{x}}=\frac{1}{G_n(y_1, \ldots, y_{n-1}, \tilde{x})} \qquad (3\text{--}14)$$

It can now be shown that every solution to the system (3–13) in D_0 is a solution of the system (3–9), and conversely. Thus, we may say that the systems (3–9) and (3–13) are equivalent in D_0.

The solutions of the system (3–9) in n variables can be found by solving the $n-1$ simultaneous equations (3–13) to determine $y_1, \ldots, y_{n-1}$ as functions of y_n. And then a single equation obtained by substituting these solutions into equation (3–14) can be solved to determine the parametric variable x as a function of y_n. In order to emphasize the connection between the autonomous system of n equations (3–9) and the system of $n-1$ equations (3–13), the former system is sometimes denoted symbolically by

$$\frac{dy_1}{G_1}=\frac{dy_2}{G_2}= \ldots =\frac{dy_n}{G_n}= dx$$

or, when finding the variable x is not important, by

$$\frac{dy_1}{G_1}=\frac{dy_2}{G_2}= \ldots =\frac{dy_n}{G_n}$$

Thus, for example, consider the autonomous system of three equations

$$\left.\begin{aligned}
\frac{dy_1}{dx} &= G_1(y_1, y_2, y_3) \equiv my_1^2 y_2 y_3^2 \\[2em]
\frac{dy_2}{dx} &= G_2(y_1, y_2, y_3) \equiv -ny_2^2 y_1 \\[2em]
\frac{dy_3}{dx} &= G_3(y_1, y_2, y_3) \equiv y_1 y_2 y_3
\end{aligned}\right\} \tag{3-15}$$

Upon putting $\tilde{x} = y_3$ we get

$$\left.\begin{aligned}
\frac{dy_1}{d\tilde{x}} &= \tilde{G}_1(y_1, y_2, \tilde{x}) \equiv my_1 \tilde{x} \\[2em]
\frac{dy_2}{d\tilde{x}} &= \tilde{G}_2(y_1, y_2, \tilde{x}) \equiv -\frac{ny_2}{\tilde{x}}
\end{aligned}\right\} \tag{3-16}$$

and x is determined by the equation

$$\frac{dx}{d\tilde{x}} = \frac{1}{\tilde{G}_3} \equiv \frac{1}{y_1 y_2 \tilde{x}} \tag{3-17}$$

The system (3–16) has the solution

$$y_1 = c_1 e^{m\tilde{x}^2/2} = c_1 e^{my_3^2/2}$$

$$y_2 = \frac{c_2}{\tilde{x}^n}$$

Finally, equation (3–17) shows that x is related to $\tilde{x}$ by

$$x = c_3 + \frac{1}{c_1 c_2} \int \tilde{x}^{n-1} \exp\left(\frac{-m\tilde{x}^2}{2}\right) d\tilde{x}$$

The interpretation of the solution of a normal system of n equations as a curve in an n-dimensional space is particularly useful when the system is autonomous. This is the case because for autonomous systems, unlike nonautonomous systems in general, there is usually only a single curve passing through each point in this space. We can show that this is the case by applying the fundamental theorem given in section 3.1 to the transformed system (3–13). This theorem applies to the $y_1, \ldots, y_{n-1}, \tilde{x}$ space in which the system (3–13) is defined. But in view of equation (3–11), we see that this is the same as the n-dimensional $y_1, \ldots, y_n$ space in which the untransformed autonomous system (3–9) is defined. Thus, suppose that the functions G_i have continuous first partial derivatives with respect to all their arguments [32] at every point in some domain D of the n-dimensional $y_1, \ldots, y_n$ space and that there is no point of D where all the G_i's simultaneously vanish. Then, because of the continuity of the G_i's, we can assert that for any point $y_1^0, \ldots, y_n^0$ of D there is a neighborhood D_0 in which at least one of the G_i's is never equal to zero. Within this neighborhood we can transform the normal system (3–9) into the system (3–13) and apply the fundamental theorem in the region D_0. Since the $\tilde{G}_i$'s have continuous partial derivatives, we can conclude that the system (3–9) has precisely one solution curve in the n-dimensional $y_1, \ldots, y_n$ space passing through the point $y_1^0, \ldots, y_n^0$. Now suppose the notation is chosen so that G_n does not vanish in D_0. Then G_n does not change sign in D_0, and equation (3–14) shows that x is a monotonic function of $\tilde{x}$ along any solution curve in D_0. Hence, we can take x as the parametric variable for the solution curve instead of $\tilde{x}$. Since this is true for every point of D, we arrive at the following conclusion: [33] *If the function* $\mathbf{G}$ *which appears in the first-order autonomous system*

$$\frac{d\mathbf{y}}{dx} = \mathbf{G}(\mathbf{y}) \tag{3–18}$$

possesses continuous partial derivatives at every point of some domain D, *and if* $|\mathbf{G}(\mathbf{y})|$ *does not vanish at any point of* D, *the system* (3–18) *has exactly one solution curve*

---

[32] Notice that in this case the existence of the partial derivatives guarantees the continuity of the functions G_i.

[33] We are using the vector notation introduced in section 3.1.

$$\mathbf{y} = \mathbf{f}(x, \mathbf{y_0}) \tag{3-19}$$

passing through each point $\mathbf{y_0}$ *of* D.

Thus, we can think of the "solution vector" (3-19) as tracing out a continuous curve in the n-dimensional $y_1, \ldots, y_n$ space, which passes through the point $\mathbf{y_0}$ when the parametric variable takes on some value, say x_0. Such "solution curves" are called *integral curves* of the system (3-18). These ideas are illustrated in figure 3-2 for the case where $n = 3$.

Notice that in order to ensure that the system (3-18) has only one integral curve passing through each point of D, it is necessary to require that $|\mathbf{G}(\mathbf{y})|$, the magnitude of $\mathbf{G}(\mathbf{y})$, does not vanish at any point of D. Points where $|\mathbf{G}(y)| = 0$ are called *critical points* of the autonomous system (3-18).

If $\mathbf{y_0}$ is a critical point of the system (3-18), this system must possess the constant solution

$$\mathbf{y} = \mathbf{y_0}$$

which is called an *equilibrium solution*.

For example, it may be verified by inspection that the solutions to the autonomous system

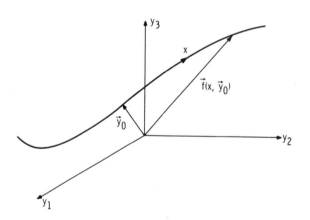

FIGURE 3-2.—Relation between solution vector and integral curve of differential equation.

$$\frac{dy_1}{dx} = my_1$$

$$\frac{dy_2}{dx} = ny_2$$

with $n > m$, are

$$y_1 = c_1 e^{mx}$$

$$y_2 = c_2 e^{nx}$$

The point $(0, 0)$ is a critical point. The integral curves of this system are the loci of points

$$(y_2)^m = k(y_1)^n$$

where $k = c_2^m / c_1^n$. These curves are shown for the case where m and n are positive integers in figure 3–3. It can be seen from this figure that there is

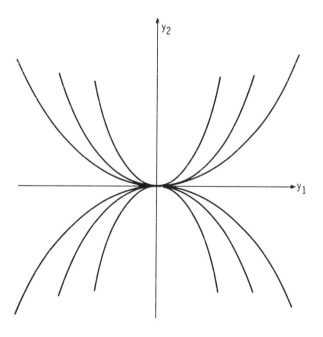

FIGURE 3–3. — Integral curves for sample problem.

exactly one integral curve passing through each point except the critical point $(0, 0)$.

A vector function which is defined at each point of an n-dimensional space is said to constitute a "vector field." Since a unique vector-valued function $\mathbf{G}(\mathbf{y})$ corresponds to each autonomous system (3–18) and conversely, we can say that each first-order autonomous system is characterized by its vector field.

Consider a portion of an integral curve of the system (3–18) which passes through a point $\mathbf{y}$ which is not a critical point and which lies in a domain D where $\mathbf{G}(\mathbf{y})$ is continuously differentiable. Let S denote the arc length measured in some definite direction along this curve. Then

$$(dS)^2 = (dy_1)^2 + (dy_2)^2 + \ldots + (dy_n)^2$$

or

$$dS = \pm dx \sqrt{\frac{d\mathbf{y}}{dx} \cdot \frac{d\mathbf{y}}{dx}}$$

But equation (3–18) shows that

$$\frac{dS}{dx} = \pm \sqrt{\mathbf{G} \cdot \mathbf{G}} = \pm |\mathbf{G}|$$

Hence, taking S as the independent variable in equation (3–18) shows that

$$\frac{d\mathbf{y}}{dS} = \frac{d\mathbf{y}}{dx}\frac{dx}{dS} = \pm \frac{\mathbf{G}}{|\mathbf{G}|}$$

However, the vector $d\mathbf{y}/dS$ is the unit tangent vector to the integral curve passing through $\mathbf{y}$; and since there is an integral curve passing through each point of D, we conclude that the vector field $\mathbf{G}(\mathbf{y})$ is tangent to the integral curves of the system (3–18), except possibly at the critical points.

When $n = 3$, we can imagine that the variable x represents time and the autonomous system (3–18) represents the steady flow of a fluid in space. At each point $\mathbf{y}$ in a certain region of this space the vector $\mathbf{G}(\mathbf{y})$ describes the velocity of the fluid at that point in both magnitude and direction. The flow is steady because its velocity depends only on position and does not vary with time. The integral curve passing through the point $\mathbf{y}_0$ may then be interpreted as the streamline followed by all the fluid particles which have passed through the point $\mathbf{y}_0$. The velocity field is tangent to the streamlines at every point except at the critical points which correspond to "stagnation points" where the fluid velocity vanishes. At these points, it is possible for two or more streamlines to meet or for a given streamline to abruptly change direction.

By again applying the results obtained in section 3.1 to the autonomous system (3–9) by means of the transformed system (3–13), we arrive at the following conclusions: Suppose that D is a region of $y_1, \ldots, y_n$ space which contains no critical points of the system (3–9) and that $\mathbf{G}$ *has continuous partial derivatives* at every point of D. Then *a function $H(\mathbf{y})$ of the vector variable $\mathbf{y}$ in D which does not depend explicitly on x is a solution of the first-order linear partial differential equation*

$$G_1(\mathbf{y})\,\frac{\partial H}{\partial y_1} + \ldots + G_n(\mathbf{y})\,\frac{\partial H}{\partial y_n} = 0 \tag{3–20}$$

if, and only if, it is an integral of the autonomous system (3–9). If, in addition, the region D is sufficiently small, the system (3–9) possesses precisely $n-1$ functionally independent integrals which are independent of x. This shows that equation (3–20) possesses $n-1$ functionally independent solutions.

The level surfaces of these integrals are hypersurfaces in the n-dimensional $y_1, \ldots, y_n$ space. The intersection of any $n-1$ of these hypersurfaces (no two of which correspond to the same functionally independent integral) is an integral curve of the system (3–9) in this space.

The partial differential equation can be written more compactly by using the vector notation introduced in section 3.1 to obtain

$$\mathbf{G} \cdot \nabla H = 0$$

In this form the equation has an immediate geometrical interpretation. As in three-dimensional space, the vector ∇H is perpendicular to the level surfaces of the function H. Hence, the differential equation (3–20) merely states that the vector field $\mathbf{G}$ is everywhere tangent to the level surfaces of its solutions. This is consistent with the facts that the integral curves of the system (3–9) lie on these level surfaces and that the vector field $\mathbf{G}$ is tangent to these integral curves.

3.3 SOLUTIONS OF THE FIRST-ORDER LINEAR PARTIAL DIFFERENTIAL EQUATION

Since equation (3–20) cannot have n functionally independent solutions, there must exist for any n solutions $H_1, \ldots, H_n$ of equation (3–20) a non-constant function ω such that

$$\omega(H_1, H_2, \ldots, H_n) = 0$$

for all $y_1, \ldots, y_n$ in some region D. Now it can be shown (ref. 6) that this equation can always be solved for H_n to obtain

$$H_n = F(H_1, H_2, \ldots, H_{n-1})$$

provided that $H_1, \ldots, H_{n-1}$ are functionally independent. But since $H_1, \ldots, H_n$ were any solutions of equation (3–20), this shows that every solution of equation (3–20) can be expressed as a function of any $n-1$ functionally independent solutions. Conversely, it is easy to verify by direct substitution that if $H_1, H_2, \ldots, H_{n-1}$ are any solutions of equation (3–20) and F is *any* continuously differentiable function of $H_1, \ldots, H_{n-1}$, then $F(H_1, H_2, \ldots, H_{n-1})$ is also a solution of equation (3–20). Thus, we have shown that *H is a solution of the linear partial differential equation (3–20) if, and only if, there is a function F such that*

$$H(\mathbf{y}) = F(H_1(\mathbf{y}), \ldots, H_{n-1}(\mathbf{y}))$$

where $H_1(y), \ldots, H_{n-1}(y)$ are any $n-1$ functionally independent integrals of the autonomous system (3–9) which do not depend on x.

Thus, the most general solution of the partial differential equation (3–20) can be found if the system (3–9) of ordinary differential equations can be solved. The differential equations (3–9) are therefore called the *characteristic equations* of the partial differential equation (3–20) and their integral curves are called *characteristic curves* of the partial differential equation.

For example, the characteristic equations of the partial differential equation

$$(y_2 - y_3) \frac{\partial H}{\partial y_1} + (y_3 - y_1) \frac{\partial H}{\partial y_2} + (y_1 - y_2) \frac{\partial H}{\partial y_3} = 0 \qquad (3\text{–}21)$$

are the equations of the system (3–8). But we have shown that this system has the two functionally independent integrals

$$H_1 = y_1 + y_2 + y_3$$

$$H_2 = y_1^2 + y_2^2 + y_3^2$$

Thus, for every differentiable function F, the function

$$H = F(H_1, H_2)$$

must be a solution of equation (3–21). This can easily be verified by substituting the relations

$$\frac{\partial H}{\partial y_j} = \frac{\partial F}{\partial H_1} + 2y_j \frac{\partial F}{\partial H_2} \qquad \text{for } j = 1, 2, 3$$

into equation (3–21).

The level surfaces of the integral H_1 are planes, and the level surfaces of the integral H_2 are the surfaces of spheres. The intersections of the surfaces $H_1 = constant$ with the surfaces $H_2 = constant$ are therefore circles. Hence, the characteristic curves of equation (3–21) are circles.

3.4 QUASI-LINEAR PARTIAL DIFFERENTIAL EQUATIONS OF THE FIRST ORDER

There is also a close connection between the quasi-linear partial differential equation

$$G_1(y_1, \ldots, y_{n-1}, f) \frac{\partial f}{\partial y_1} + \ldots + G_{n-1}(y_1, \ldots, y_{n-1}, f) \frac{\partial f}{\partial y_{n-1}}$$

$$= G_n(y_1, \ldots, y_{n-1}, f) \qquad (3\text{--}22)$$

and the linear partial differential equation (3–20) and, therefore, also between this equation and the autonomous system (3–9). In order to show this, let

$$H = h(y_1, \ldots, y_n)$$

be any solution of equation (3–20) such that $\partial H/\partial y_n$ is not identically zero.[34] Then the implicit function theorem shows that there is a neighborhood D of any point $\mathbf{y}_0$ for which

$$\frac{\partial H}{\partial y_n} \neq 0 \tag{3–23}$$

such that the equation

$$h(y_1, \ldots, y_n) = c_1 \tag{3–24}$$

has a solution

$$y_n = f(y_1, \ldots, y_{n-1}, c_1) \tag{3–25}$$

in D. This means that when equation (3–25) is substituted into equation (3–24), the resulting expression is identically constant for all values of $y_1, \ldots, y_{n-1}$. Hence, upon differentiating this expression successively with respect to $y_1, \ldots, y_{n-1}$, we obtain

$$\frac{\partial H}{\partial y_1} + \frac{\partial H}{\partial y_n} \frac{\partial f}{\partial y_1} = 0$$

$$\vdots \qquad \vdots$$

$$\frac{\partial H}{\partial y_{n-1}} + \frac{\partial H}{\partial y_n} \frac{\partial f}{\partial y_{n-1}} = 0$$

But when these relations are substituted into equation (3–20), we find that

$$\frac{\partial H}{\partial y_n}\left(G_n - G_1\frac{\partial f}{\partial y_1} - \ldots - G_{n-1}\frac{\partial f}{\partial y_{n-1}}\right) = 0$$

And, in view of equation (3–23), this shows that equation (3–25) is a solution of equation (3–22).

---

[34] This means that H is not independent of y_n.

Now there are $n-1$ functionally independent integrals $H_1, \ldots, H_{n-1}$ of the system (3–9), which are therefore functionally independent solutions of equation (3–20). If there is some point $\mathbf{y}$ at which these integrals all satisfy condition (3–23), we can find at least $n-1$ functions

$$\left.\begin{array}{c} y_n = f_1(y_1, \ldots, y_{n-1}, c_1) \\ \cdot \\ \cdot \\ \cdot \\ y_n = f_{n-1}(y_1, \ldots, y_{n-1}, c_{n-1}) \end{array}\right\} \qquad (3\text{–}26)$$

which, respectively, satisfy the equations

$$\left.\begin{array}{c} H_1 = h_1(y_1, \ldots, y_n) = c_1 \\ \cdot \\ \cdot \\ \cdot \\ H_{n-1} = h_{n-1}(y_1, \ldots, y_n) = c_{n-1} \end{array}\right\} \qquad (3\text{–}27)$$

in some neighborhood D of the point $\mathbf{y}$. Therefore, each of the functions (3–26) is a solution to equation (3–22) in D. Since equations (3–26) determine explicitly the same surfaces as equations (3–27) determine implicitly and since these surfaces determine the integral curves of the system (3–9), the equations of this system are also called the characteristic equations of equation (3–22).

For example, the partial differential equation

$$(y_2 - f) \frac{\partial f}{\partial y_1} + (f - y_1) \frac{\partial f}{\partial y_2} = y_1 - y_2 \qquad (3\text{–}28)$$

has the same characteristic equations as the linear equation in the preceding example, namely equations (3–8). These equations, as we have seen, have the two functionally independent integrals

$$H_1 = y_1 + y_2 + y_3$$

$$H_2 = y_1^2 + y_2^2 + y_3^2$$

And equation (3–28) therefore has the solutions

$$f_1 = c_1 - (y_1 + y_2)$$

$$f_2 = \pm \sqrt{c_2^2 - (y_1^2 + y_2^2)}$$

CHAPTER 4

Elementary Methods for Second-Order Equations

In this chapter several methods for obtaining solutions to second-order differential equations are presented. Even though each of these methods applies to a relatively narrow class of equations, they are still of sufficient generality to be useful in practice. However, unlike the methods presented in chapter 2 for solving first-order equations, most of the techniques presented in this chapter will not by themselves yield solutions to equations. They are, in fact, methods for transforming certain types of equations into simpler equations whose solutions can hopefully be found by known methods. We have already found a number of methods for solving first-order equations. The only other general class of equations for which it is possible to find numerous solutions is the class of linear equations introduced in chapter 1. Therefore, in section 4.1 we present some techniques for reducing second-order equations to equations of the first order. And in section 4.2 we present techniques which transform certain types of nonlinear equations into linear equations. Finally, in section 4.3 we present a number of unrelated methods. Although many of the methods presented in this chapter are applicable, with some obvious modification, to equations of higher order, we shall, for simplicity, limit the discussion to second-order equations.

As indicated in chapter 1, the most general second-order differential equation is an equation of the form

$$F\left(\frac{d^2y}{dx^2}, \frac{dy}{dx}, y, x\right) = 0 \tag{4–1}$$

whereas, the most general normal second-order equation is of the form

$$\frac{d^2y}{dx^2} = G\left(\frac{dy}{dx}, y, x\right) \tag{4-2}$$

A general solution of equation (4–1) or of equation (4–2), if it exists, will involve two arbitrary constants. As in the case of first-order equations, there may be any number of general solutions to these equations. In addition, equation (4–1) may possess singular solutions. These solutions are discussed in section 1.5. The methods presented in that section should be used to find any singular solutions which may be present since such solutions can be important in physical problems.

Finally, the second-order linear equation is an equation of the form

$$p_0(x)\frac{d^2y}{dx^2} + p_1(x)\frac{dy}{dx} + p_2(x)y = p_3(x) \tag{4-3}$$

We shall assume the coefficients are continuous. Upon dividing through by the leading coefficient $p_0(x)$, we obtain the (essentially) normal form

$$\frac{d^2y}{dx^2} + p(x)\frac{dy}{dx} + q(x)y = r(x) \tag{4-4}$$

This differential equation is equivalent to equation (4–3) in any interval in which $p_0(x)$ does not vanish.

4.1 EQUATIONS WHICH ARE REDUCIBLE TO FIRST-ORDER EQUATIONS

In this section a number of techniques, which can be used to reduce certain types of equations to equations of the first order, are described.

4.1.1 Dependent Variable Missing

When a second-order differential equation does not explicitly contain the dependent variable y, it must be of the form

$$F\left(\frac{d^2y}{dx^2}, \frac{dy}{dx}, x\right) = 0 \tag{4-5}$$

But this equation can be written as a first-order differential equation

$$F\left(\frac{dp}{dx}, p, x\right) = 0$$

for the quantity p defined by

$$p \equiv \frac{dy}{dx} \tag{4-6}$$

It may or may not be possible to solve this first-order equation in any given case. However, suppose that it can be solved and that its integral is

$$f(x, p) = \text{constant} = C_1 \tag{4-7}$$

At this stage, there are two possible ways of proceeding, depending on whether it is easier to find an explicit formula for p as a function of x which solves[35] equation (4–7) or whether it is easier to find an explicit formula for x as a function of p.

First, suppose that the former case occurs. Thus, we can find an explicit formula

$$p = g(x, C_1)$$

for a solution to equation (4–7). Then, in view of equation (4–6), this equation can be immediately integrated to obtain a general solution

$$y = \int g(x, C_1)\,dx + C_2$$

of equation (4–5).

Next suppose that it is easier to find an explicit formula

$$x = h(p, C_1) \tag{4-8}$$

for a solution to equation (4–7). Substituting equation (4–8) into the relation

$$p = \frac{dy}{dp}\frac{dp}{dx}$$

---

shows that

$$\frac{dy}{dp} = p\frac{dx}{dp} = p\frac{dh}{dp}$$

Hence,

$$y = \int p\frac{dh}{dp}\,dp + C_2$$

And, upon integrating this by parts, we obtain the equation

$$y = ph(p, C_1) - \int h(p, C_1)\,dp + C_2 \qquad (4\text{-}9)$$

which together with equation (4–8) determines the solution y of equation (4–5) parametrically as a function of x (with p being the parametric variable). This parametric function involves two arbitrary constants.

For example, the equation of the curve followed in the pursuit of a prey which moves along the y-axis (ref. 9) is

$$K\sqrt{1+y'^2} = (a-x)y''$$

where K is the ratio of the velocity of the prey to the velocity of the pursuer. Put $y' = p$ to obtain the first-order separable equation

$$K\sqrt{1+p^2} = (a-x)\frac{dp}{dx}$$

which can easily be solved (section 2.3) to obtain

$$\frac{dy}{dx} = p = -\frac{1}{2}\left[C_1(a-x)^K - \frac{1}{C_1}(a-x)^{-K}\right]$$

If $K \neq 1$, this equation becomes, upon integration,

$$y = \frac{1}{2}\left[\frac{C_1}{1+K}(a-x)^{K+1} + \frac{1}{C_1(K-1)}(a-x)^{1-K}\right] + C_2$$

(If $K = 1$, the integration yields a logarithmic term.)

4.1.2 Independent Variable Missing

The general form of the second-order equation in which the independent variable does not appear explicitly is

$$F\left(\frac{d^2y}{dx^2}, \frac{dy}{dx}, y\right) = 0 \qquad (4\text{–}10)$$

We again define p by

$$p \equiv \frac{dy}{dx} \qquad (4\text{–}11)$$

Then substituting this together with the relation

$$\frac{d^2y}{dx^2} = \frac{dp}{dx} = \frac{dp}{dy}\frac{dy}{dx} = \frac{dp}{dy}p \qquad (4\text{–}12)$$

into equation (4–10) shows that p is determined by the first-order differential equation

$$F\left(\frac{dp}{dy}p, p, y\right) = 0$$

Suppose that this equation can be solved and its integral is

$$f(p, y) = C_1 = \text{constant} \qquad (4\text{–}13)$$

It is again possible to proceed in two different ways. First, suppose that it is easier to find a formula for the solution $p = g(y, C_1)$ of equation (4–13) for p as a function of y. Then combining this with equation (4–11) and integrating provides a general solution

$$\int \frac{dy}{g(y, C_1)} + C_2 = x$$

of equation (4–10).

On the other hand, if it is easier to find a formula

$$y = h(p, C_1) \qquad (4\text{–}14)$$

for the solution of equation (4–13) which expresses y as a function of p, equation (4–12) shows that

$$\frac{dx}{dp}=\frac{1}{p}\frac{dy}{dp}=\frac{1}{p}\frac{dh}{dp}$$

But this equation can be integrated by parts to obtain the equation

$$x=C_2+\frac{1}{p}h(p,\,C_1)+\int\frac{1}{p^2}h(p,\,C_1)dp \qquad (4\text{--}15)$$

which, together with equation (4–14), determines the solution to equation (4–10) parametrically.

For example, the differential equation

$$y\frac{d^2y}{dx^2}+\nu\left(\frac{dy}{dx}\right)^2=0 \qquad (4\text{--}16)$$

occurs in the field of fluid dynamics.[36] Substituting equations (4–11) and (4–12) into this equation yields the first-order equation

$$yp\frac{dp}{dy}+\nu p^2=0$$

Since this equation is separable, it can be integrated to obtain $p=C_1/y^\nu$. Hence, combining this with equation (4–11) and integrating shows that

$$\frac{y^{\nu+1}}{\nu+1}=C_1x+C_2$$

is a solution of equation (4–16).

Notice that, when A, B, C, and D are constants, the differential equation

$$y''=A+By+Cy^2+Dy^3$$

---

[36] More specifically, it governs the velocity field in a boundary layer in the neighborhood of the separation point.

does not explicitly contain the dependent variable x. When the procedure just described is applied to this equation, we encounter integrals which usually cannot be evaluated in terms of elementary functions. These integrals, however, can always be expressed in terms of *elliptic functions*. The differential equation is therefore called the *elliptic* equation. For more details concerning the procedures involved in evaluating these integrals and the properties of elliptic functions, the reader is referred to reference 10 (chapter 6 and section 11 of chapter 7).

Finally, notice that a linear homogeneous equation does not contain the independent variable explicitly if, and only if, it has constant coefficients.[37]

4.1.3 Homogeneous Equations

The general definition of a homogeneous function of degree k has been given in section 2.3. We shall now show that if the second-order differential equation (4–1) has certain homogeneity properties, it can be solved by elementary means.

4.1.3.1 *Equations homogeneous in the dependent variable and its derivatives.*—First, suppose that equation (4–1) is homogeneous of degree k in the variables y, y', and y''; that is, it satisfies the homogeneity condition

$$F(ty'', ty', ty, x) = t^k F(y'', y', y, x) \qquad (4\text{–}17)$$

for any number t. Hence, if we put $t = 1/y$, equation (4–17) becomes

$$F\left(\frac{y''}{y}, \frac{y'}{y}, 1, x\right) = \frac{1}{y^k} F(y'', y', y, x)$$

which shows that the differential equation (4–1) can be written in the form

$$F\left(\frac{y''}{y}, \frac{y'}{y}, 1, x\right) = 0 \qquad (4\text{–}18)$$

by factoring out y^k. In order to reduce the order of this equation, define the function $u(x)$ by

---

[37] We assume that the reader knows how to solve linear equations with constant coefficients.

$$u = \frac{y'}{y} \tag{4-19}$$

Then

$$\frac{y''}{y} = u' + u^2$$

Upon substituting this, together with equation (4–19), into equation (4–18) we obtain the first-order equation

$$F(u' + u^2, u, 1, x) = 0 \tag{4-20}$$

Suppose this equation can be solved and its integral is

$$f(u, x) = C_1 \tag{4-21}$$

Then solve equation (4–21) for u as a function of x to obtain

$$u = g(x, C_1) \tag{4-22}$$

But substituting this into equation (4–19) and integrating shows that the original equation (4–1) has the solution

$$\ln y = \int g(x, C_1)\,dx + C_2 \tag{4-23}$$

If the differential equation (4–1) is linear, the first-order equation (4–20) obtained by this method will be a Riccati equation which will usually be more difficult to solve than the original equation.

As an example of the method, consider the equation

$$F(y'', y', y, x) = yy'' - y'^2 + y^2x = 0 \tag{4-24}$$

and replace y, y', and y'' by ty, ty', and ty'', respectively, to get

$$(ty)(ty'') - (ty')^2 + (ty)^2x = t^2(yy'' - y'^2 + y^2x) = t^2 F(y'', y', y, x)$$

This shows that F is homogeneous of degree 2 in y and its derivative. Hence, upon introducing the transformation (4–19), the differential equation (4–24) becomes

$$u' + u^2 - u^2 + x = u' + x = 0$$

But this equation can be immediately integrated to obtain

$$u + \frac{x^2}{2} = C_1$$

Hence,

$$\ln y = \int \left(C_1 - \frac{x^2}{2} \right) dx + C_2$$

The solution to equation (4–24) is, therefore, given by

$$y = C_3 \exp \left(C_1 x - \frac{x^3}{6} \right)$$

where we have put $C_3 = e^{C_2}$.

4.1.3.2 *Isobaric equation.* — The isobaric equation of the first order in normal form was encountered in section 2.3. In the general case, the function

$$F \left(\frac{d^m y}{dx^m}, \frac{d^{m-1} y}{dx^{m-1}}, \ \ldots, \ \frac{dy}{dx}, y, x \right)$$

is said to be *isobaric* if there exist numbers k and l such that

$$F \left(t^{l-m} \frac{d^m y}{dx^m}, t^{l-m+1} \frac{d^{m-1} y}{dx^{m-1}}, \ \ldots, \ t^{l-1} \frac{dy}{dx}, t^l y, tx \right)$$

$$= t^k F \left(\frac{d^m y}{dx^m}, \frac{d^{m-1} y}{dx^{m-1}}, \ \ldots, \ \frac{dy}{dx}, y, x \right) \quad (4\text{–}25)$$

for all values of t. And the differential equation

$$F [y^{(m)}, y^{(m-1)}, \ \ldots, y', y, x] = 0$$

is said to be *isobaric* if the function F is isobaric. It is easy to see that the differential equation (2–17) is of this type (for $m=1$). For the second-order differential equation

$$F(y'', y', y, x)=0 \qquad (4\text{–}26)$$

the condition (4–25) becomes

$$F(t^{l-2}y'', t^{l-1}y', t^l y, tx)=t^k F(y'', y', y, x) \qquad (4\text{–}27)$$

Thus, in particular, upon replacing t by $1/x$, equation (4–27) becomes

$$F\left(\frac{y''}{x^{l-2}}, \frac{y'}{x^{l-1}}, \frac{y}{x^l}, 1\right)=\frac{1}{x^k} F(y'', y', y, x)=0$$

And this shows that equation (4–26) can be written in the form

$$F\left(\frac{y''}{x^{l-2}}, \frac{y}{x^{l-1}}, \frac{y}{x^l}, 1\right)=0 \qquad (4\text{–}28)$$

However, when we introduce the new variables

$$u=x^{-l}y \qquad (4\text{–}29)$$

and

$$\xi=\ln x \qquad (4\text{–}30)$$

into this equation, we obtain an equation

$$F\left[\frac{d^2u}{d\xi^2}+(2l-1)\frac{du}{d\xi}+l(l-1)u, \frac{du}{d\xi}+lu, u, 1\right]=0 \qquad (4\text{–}31)$$

in which the independent variable ξ does not appear explicitly. Since this equation can always be reduced to a first-order equation by the methods of section 4.1.2, it follows that the isobaric equation of the second order can also be reduced to such an equation.

The homogeneous linear isobaric equation of the second order is an equation of the form

$$x^2 y'' + p_0 x y' + q_0 y = 0 \qquad (4\text{--}32)$$

where p_0 and q_0 are constants. It is known as the homogeneous *Euler equation*. Since this equation satisfies condition (4–27) with $l = k = 0$, the change of variables $u = y$, $\xi = \ln x$ will transform it into the equation

$$\frac{d^2 u}{d\xi^2} + (p_0 - 1)\frac{du}{d\xi} + q_0 u = 0$$

The independent variable does not appear explicitly in this equation and the coefficients are constants.[38]

This equation, therefore, has two solutions of the form $u = e^{C\xi}$, one for each of the roots of the equation.

$$C^2 + (p_0 - 1)C + q_0 = 0$$

It is easy to see that the general Euler equation

$$x^2 y'' + p_0 x y' + q_0 y = r(x)$$

can also be transformed into an equation with constant coefficients by the change of variables

$$u = y \qquad \xi = \ln x$$

A less elementary example of an isobaric equation is provided by the equation

$$F(y'', y', y, x) = yy'' - \frac{3}{4}y'^2 + \left(\frac{y}{x}\right)^2 = 0 \qquad (4\text{--}33)$$

It is easy to verify that, when the function F is given by equation (4–33), the condition (4–27) will be satisfied for any value of l provided that $k = 2l - 2$. Hence, upon introducing the change of variables

---

[38]Which is as it should be since the equation is linear.

$$u = \frac{y}{x^l} \\ \\ \xi = \ln x \Bigg\} \tag{4-34}$$

the differential equation (4–33) is transformed into the equation

$$u \frac{d^2u}{d\xi^2} + \left(\frac{l}{2} - 1\right) u \frac{du}{d\xi} - \frac{3}{4}\left(\frac{du}{d\xi}\right)^2 + \left(\frac{l}{2} - 1\right)^2 u^2 = 0 \tag{4-35}$$

Since in this case any value of l can be used, we choose its value to simplify equation (4–35). This is accomplished by setting $l = 2$, whereupon equation (4–35) becomes

$$u \frac{d^2u}{d\xi^2} - \frac{3}{4}\left(\frac{du}{d\xi}\right)^2 = 0$$

But this is a special case of equation (4–16) with $\nu = -3/4$ and, therefore, has the solution

$$4u^{1/4} = C_1 \xi + C_2$$

Hence, by using equations (4–34) and (4–35), the solution to equation (4–33) is found to be

$$y = x^2 \left(\frac{C_1 \ln x + C_2}{4}\right)^4$$

4.1.4 Method of Variation of Parameters

This method applies only to linear equations. We have seen in chapter 1 that the homogeneous equation (associated with eq. (4–4))

$$y'' + p(x)y' + q(x)y = 0 \tag{4-36}$$

possesses exactly two linearly independent solutions in any interval in which its coefficients are continuous, and that the general solution of the nonhomogeneous equation (4–4) in this interval is the sum of any particular solution of that equation plus an arbitrary linear combination of two linearly independent solutions of (4–36).

We now suppose that by some means a nontrivial solution y_H of the associated homogeneous equation (4–36) has been found. Substituting

$$y = v y_H$$

(where v is a function to be determined subsequently) into equation (4–4) and using the fact that y_H satisfies the homogeneous equation (4–36) to simplify the result shows that

$$y_H v'' + (2y_H' + p y_H) v' = r$$

But this is a first-order linear equation for v'. It can therefore be solved by the methods of section 2.3, to obtain

$$y_H^2 v' e^{\int p \, dx} = \int r(x) y_H(x) e^{\int p(x) dx} \, dx + C_1$$

When this equation is solved for v' and integrated, we obtain an expression for v (in terms of known functions) which contains two arbitrary constants. Hence, substituting this expression into $y = y_H v$ gives the general solution to equation (4–4).

For example, it is easy to find by inspection that $y_H = x$ is a homogeneous solution of the equation

$$y'' - \frac{2x}{1+x^2} y' + \frac{2}{1+x^2} y = \frac{x^3 + 3x}{1+x^2} \tag{4-37}$$

Hence, substituting $y = xv$ into this equation shows that v' satisfies the first-order linear equation

$$v'' + \frac{2}{x(x^2+1)} - v' = \frac{x^2+3}{x^2+1}$$

which can easily be integrated to obtain

$$\frac{x^2 v'}{x^2+1} = \frac{x^3}{x^2+1} + C_1$$

But solving this equation for v' and integrating shows that v is given by

$$v = \frac{x^2}{2} + C_1\left(x - \frac{1}{x}\right) + C_2$$

Therefore, the general solution to equation (4–37) is

$$y = xv = \frac{x^3}{2} + C_1(x^2-1) + C_2 x$$

4.1.5 Equations Invariant Under a Transformation Group

The group-theory method for reducing the order of equations is a generalization of all the techniques previously discussed in this section. However, this technique cannot be applied routinely since, as will be seen, it is necessary to find a group under which a given equation is invariant and there is no constructive procedure for accomplishing this. The method is more useful for working backwards to find general methods for solving particular classes of equations by starting with a given group and testing equations for invariance under this group in much the same way that equations are tested to see if they are homogeneous.

A *single parameter Lie group*[39] in two dimensions is a family of coordinate transformations

$$\left.\begin{array}{l} x_1 = f(x, y; \alpha) \\[6pt] y_1 = g(x, y; \alpha) \end{array}\right\} \qquad (4–38)$$

in which the members of the family are individuated by the values of the parameter α. In addition, the family must contain an identity transformation which, without loss of generality, we can associate with the value $\alpha = 0$. Thus

---

[39]Various continuity hypotheses, which will not be presented here, are required for a rigorous treatment of the theory.

$$x = f(x, y; 0)$$

$$y = g(x, y; 0)$$

Consider an infinitely differentiable but otherwise arbitrary function of the coordinates $F(x, y)$. If the point (x, y) is transformed by the group (4–38) into the point (x_1, y_1) for a particular value of α, the value of F at (x_1, y_1) is given by the Taylor series whose leading terms are

$$F(x_1, y_1) = F(x, y) + \alpha \left(\frac{\partial f}{\partial \alpha} \frac{\partial}{\partial x} + \frac{\partial g}{\partial \alpha} \frac{\partial}{\partial y} \right)_{\alpha = 0} F(x, y) + \cdots$$

$$= F(x, y) + \alpha \left[\xi(x, y) \frac{\partial}{\partial x} + \eta(x, y) \frac{\partial}{\partial y} \right] F(x, y) + \cdots \tag{4–39}$$

where we have put

$$\xi(x, y) \equiv \left(\frac{\partial f}{\partial \alpha} \right)_{\alpha = 0} \quad \text{and} \quad \eta(x, y) \equiv \left(\frac{\partial g}{\partial \alpha} \right)_{\alpha = 0} \tag{4–40}$$

In particular, if F is taken successively as x and y, then

$$\left. \begin{aligned} x_1 &= x + \alpha \xi(x, y) + \cdots \\ y_1 &= y + \alpha \eta(x, y) + \cdots \end{aligned} \right\} \tag{4–41}$$

When α is infinitesimally small, the transformation (4–41) differs only infinitesimally from the identity transformations; therefore, ξ and η are referred to as the *infinitesimal transformations* of the group. The operator

$$U \equiv \xi(x, y) \frac{\partial}{\partial x} + \eta(x, y) \frac{\partial}{\partial y}$$

is called the *infinitesimal operator* of the group. By using this operator notation, the Taylor series (4–39) can be expressed as

$$F(x_1, y_1) = F(x, y) + \alpha UF + \frac{1}{2} \alpha^2 U^2 F + \cdots$$

87

But this series can be formally summed to obtain

$$F(x_1, y_1) = e^{\alpha U} F(x, y)$$

In the particular cases where F is taken to be the coordinates themselves, this becomes

$$x_1 = e^{\alpha U} x$$

$$y_1 = e^{\alpha U} y$$

These equations are equivalent to the equations (4–38) which define the original finite transformation. This shows that the finite transformation is completely determined by the infinitesimal transformations.

In order to illustrate these ideas, consider the magnification group

$$\left.\begin{aligned} x_1 &= f(x, y; \alpha) \equiv e^{\alpha} x \\ y_1 &= g(x, y; \alpha) \equiv e^{\alpha} y \end{aligned}\right\} \tag{4–42}$$

It is apparent that we obtain the identity transformation when $\alpha = 0$. Inserting the equations of this group into equations (4–40) shows that the infinitesimal transformations of this group are

$$\xi(x, y) = \left(\frac{\partial f}{\partial \alpha}\right)_{\alpha = 0} = x$$

$$\eta(x, y) = \left(\frac{\partial g}{\partial \alpha}\right)_{\alpha = 0} = y$$

and therefore that the infinitesimal operator is

$$U \equiv x \frac{\partial}{\partial x} + y \frac{\partial}{\partial y}$$

Applying this operator repeatedly to the coordinates yields

$$Ux=x \qquad Uy=y$$
$$U^2x=x \qquad U^2y=y$$
$$U^3x=x \qquad U^3y=y$$

$$.$$
$$.$$
$$.$$

Hence,

$$x_1 = x + \alpha x + \frac{1}{2}\alpha^2 x + \frac{1}{3!}\alpha^3 x + \dots$$

$$y_1 = y + \alpha y + \frac{1}{2}\alpha^2 y + \frac{1}{3!}\alpha^3 y + \dots$$

And upon summing these series, we obtain the original group

$$x_1 = e^\alpha x \qquad \qquad y_1 = e^\alpha y$$

The differential equation

$$G(x, y, y', y'') = 0 \qquad\qquad (4\text{--}43)$$

is said to be invariant under the group (4–38) if introducing the new variables x_1 and y_1, given by equations (4–38), into this equation leads to the equation

$$G(x_1, y_1, y'_1, y''_1) = 0$$

This means that the change of variable given in equations (4–38) does not alter the form of the differential equation (4–43).

For example, the differential equation

$$G(x, y, y', y'') \equiv xy'' - F\left(\frac{y}{x}, y'\right) = 0 \qquad\qquad (4\text{--}44)$$

is invariant under the magnification group (4–42) since it follows from equations (4–42) that

$$\frac{y}{x}=\frac{y_1}{x_1}, \quad \frac{dy}{dx}=\frac{dy_1}{dx_1}, \quad x\frac{d^2y}{dx^2}=x_1\frac{d^2y_1}{dx_1^2}$$

We state without proof that a necessary condition for the differential equation (4–43) to be invariant under the group (4–38) is that

$$U_0G=0 \tag{4–45}$$

where the operator U_0 is defined by

$$U_0\equiv\xi\frac{\partial}{\partial x}+\eta\frac{\partial}{\partial y}+\zeta\frac{\partial}{\partial y'}+\chi\frac{\partial}{\partial y''}$$

with

$$\zeta=\frac{d\eta}{dx}-y'\frac{d\xi}{dx}$$

and

$$\chi=\frac{d^2\eta}{dx^2}-y''\frac{d\xi}{dx}-y'\frac{d^2\xi}{dx^2}$$

In performing the partial derivatives we treat x, y, y', and y'' as independent variables; whereas, in performing the total differentiations we treat these quantities as functions of x.

For example, since we have shown that

$$\xi=x \qquad \eta=y$$

for the magnification group (4–42), it follows that

$$\zeta=y'-y'\frac{dx}{dx}=0$$

$$\chi=y''-y''\frac{dx}{dx}-y'\cdot0=0$$

for this group. The operator U_0 is, therefore, given by

$$U_0 = x \frac{\partial}{\partial x} + y \frac{\partial}{\partial y} \qquad (4\text{-}46)$$

But since

$$\frac{\partial}{\partial x} F\left(\frac{y}{x}, y\right) = \frac{y}{x} \frac{\partial}{\partial y} F\left(\frac{y}{x}, y\right)$$

applying the operator (4–46) to equation (4–44) shows that

$$U_0 G = -y \frac{\partial F}{\partial y} + y \frac{\partial F}{\partial y} = 0$$

which is consistent with the result obtained in the previous example.

We have shown in chapter 3 that the system[40]

$$\frac{dx}{\xi(x,y)} = \frac{dy}{\eta(x,y)} = \frac{dp}{\zeta(x,y,p)} \qquad (4\text{-}47)$$

has two functionally independent integrals, $u(x,y)$ and $v(x,y,p)$, one of which can be chosen independent of p. The following result allows us to apply the ideas of group theory to reduce the order of a differential equation: *If the differential equation (4–43) is invariant under the group (4–38), then the equation obtained by introducing the integrals u and v as new variables in this equation is of the first order.*

Thus, we have seen in the preceding examples that equation (4–44) is invariant under the magnification group (4–42) and that, for this group,

$$\xi = x \qquad \eta = y \qquad \zeta = 0 \qquad \chi = 0$$

Hence, the system (4–47) becomes

$$\frac{dx}{x} = \frac{dy}{y} = \frac{dp}{0}$$

---

[40] We have written $p = y'$.

which has the two functionally independent integrals

$$u = \frac{y}{x} \quad \text{and} \quad v = p$$

But since

$$y'' = \frac{dp}{dx} = \frac{dv}{dx} = \frac{dv}{du}\frac{du}{dx} = \frac{dv}{du}\left(\frac{p}{x} - \frac{y}{x^2}\right)$$

we see that

$$xy'' = \frac{dv}{du}(v - u)$$

Hence, upon introducing the new variables u and v into equation (4–44), we get the first-order equation

$$\frac{dv}{du} = \frac{F(u, v)}{u - v}$$

It is easy to see that equation (4–44) is the general isobaric equation with $l = 1$. For this case the group-theory method is the same as the method given in section 4.1.3.2. In fact, all the methods of solution given up to now in this chapter are equivalent to the group-theory method when used in conjunction with certain well-known groups.

For example, the equation with the dependent variable missing, treated in section 4.1.1, is invariant under the translation group

$$x_1 = x$$
$$y_1 = y + \alpha$$

And the equation with independent variable missing, treated in section 4.1.2, is invariant under the translation group

$$x_1 = x + \alpha$$
$$y_1 = y$$

The isobaric equation with $l=0$, treated in section 4.1.3.2, is invariant under the affine group

$$x_1 = e^{\alpha} x$$

$$y_1 = y$$

And the homogeneous equation treated in section 4.1.3.1 is invariant under the affine group

$$x_1 = x$$

$$y_1 = e^{\alpha} y$$

The general linear equation discussed in section 4.1.4 is invariant under the nonuniform-distortion group

$$x_1 = x$$

$$y_1 = y + \alpha \phi(x)$$

where ϕ is any homogeneous solution of the equation.

In each of these cases the group-theory method is entirely equivalent to the method already introduced.

4.1.6 Exact Equations of the Second Order

We introduced the exact equation of the first order in section 2.2. We shall now extend the ideas presented therein to the second-order differential equation

$$F(y'', y', y, x) = 0 \qquad (4{-}48)$$

We say that the differential equation (4–48) is *exact* if there exists a function $\phi(y', y, x)$ such that

$$F = \frac{d\phi}{dx} \qquad (4{-}49)$$

Suppose that such a function ϕ exists. Then every solution of equation (4–48) must also be a solution of the first-order differential equation

$$\phi(y', y, x) = C_1 \qquad (4\text{–}50)$$

for some constant C_1. Hence, the problem of solving equation (4–48) can be replaced by the problem of solving the first-order equation (4–50). We can therefore say that equation (4–48) has been reduced to the first-order equation (4–50). Upon recalling the definition given in section 1.3 we see that ϕ is a (first) integral of equation (4–48).

It follows from the chain rule that

$$\frac{d\phi}{dx} = \frac{\partial \phi}{\partial x} + \frac{\partial \phi}{\partial y} y' + \frac{\partial \phi}{\partial y'} y'' \qquad (4\text{–}51)$$

where the variables x, y, and y' are treated as independent in forming the partial derivatives. Hence, it follows from equation (4–49) and the fact that ϕ does not depend on y'' that the differential equation (4–48) must be an equation of the form

$$F(y'', y', y, x) = f(x, y, y')y'' + g(x, y, y') = 0 \qquad (4\text{–}52)$$

where [41]

$$f(x, y, p) = \phi_p \qquad (4\text{–}53)$$

$$g(x, y, p) = \phi_x + p\phi_y \qquad (4\text{–}54)$$

It is now easy to verify by substituting in equations (4–53) and (4–54) that f and g satisfy the conditions

$$\left. \begin{array}{l} f_{xx} + 2pf_{xy} + p^2 f_{yy} = g_{xp} + pg_{yp} - g_y \\[2em] f_{xp} + pf_{yp} + 2f_y = g_{pp} \end{array} \right\} \qquad (4\text{–}55)$$

---

[41] As usual, we have put $p = y'$. The subscript notation for partial derivatives is being used.

We have therefore shown that, if the general second-order equation (4–48) is exact, it must be of the first degree and, hence, of the form (4–52). And the functions f and g in this equation must satisfy conditions (4–55). Then the equation will have a first integral ϕ (and therefore can be reduced to a first-order equation) which can be found by integrating equation (4–53) with respect to p (at constant x and y) to obtain

$$\phi = \int f(x, y, p)\,dp + h(x, y) \qquad (4\text{–}56)$$

where h is an arbitrary function which arises from the integration. Now if the differential equation is exact, it will always be possible to determine the function h so that, when equation (4–56) is substituted into equation (4–54), the latter equation will be identically satisfied.

In order to illustrate the method, consider the equation

$$F(y'', y', x) \equiv xyy'' + xy'^2 + yy' = 0 \qquad (4\text{–}57)$$

For this equation the functions f and g (in eq. (4–52)) are given by

$$\left.\begin{aligned} f &= xy \\[2mm] g &= xy'^2 + yy' = xp^2 + yp \end{aligned}\right\} \qquad (4\text{–}58)$$

It is easy to verify that these relations satisfy conditions (4–55). Hence, equation (4–57) is an exact equation.

Substituting the first of equations (4–58) into equation (4–56) shows that

$$\phi = pxy + h(x, y)$$

and substituting this and the second equation (4–58) into equation (4–54) shows that

$$xp^2 + yp = yp + h_x + p^2x + ph_y$$

But this equation is satisfied by taking $h = \text{constant}$. Hence, equation (4–57) has the first integral

$$\phi = pxy = y'xy = C_1 \tag{4-59}$$

where C_1 is a constant. But this first-order equation is separable and can be integrated immediately to obtain the general solution

$$\frac{y^2}{2} = C_1 \ln x + C_2$$

of equation (4-57).

Sometimes a first integral of an exact equation can be found by inspection simply by collecting terms and writing the equation in the form (4-49). Thus, since

$$\frac{d}{dx} yy' = y'^2 + yy''$$

the differential equation (4-57) can also be written as

$$F(y', y, x) = x \frac{d}{dx} (yy') + (yy') = 0$$

or

$$F(y', y, x) = \frac{d}{dx} xyy' = 0$$

Hence,

$$\phi = xyy'$$

which is the same as equation (4-59).

The general linear equation (4-3) is of the form (4-52) with

$$f(x, y, y') = p_0(x)$$

and

$$g(x, y, y') = p_1(x)y' + p_2(x)y - p_3(x)$$

The second condition (4–55) for exactness is automatically satisfied; and the first condition is satisfied if, and only if,

$$\frac{d^2p_0}{dx^2} - \frac{dp_1}{dx} + p_2 = 0 \tag{4-60}$$

And when this is the case, equation (4–56) becomes

$$\phi = p_0(x)p + h(x, y)$$

But substituting this into equation (4–54) shows that

$$\left(\frac{dp_0}{dx} - p_1 + \frac{\partial h}{\partial y}\right) p = p_2 y - \frac{\partial h}{\partial x} - p_3$$

Hence, it follows from equation (4–60) that

$$\frac{\partial}{\partial y}\left[\left(\frac{dp_0}{dx} - p_1\right) y + h\right] p + \frac{\partial}{\partial x}\left[\left(\frac{dp_0}{dx} - p_1\right) y + h\right] + p_3 = 0$$

It is easy to see that this equation will be satisfied when h is given by

$$h = -\left(\frac{dp_0}{dx} - p_1\right) y - \int p_3(x)\,dx$$

Hence, we have shown that the first integral of the general, linear, exact equation is

$$\phi = p_0(x)y' - \left(\frac{dp_0}{dx} - p_1\right) y - \int p_3(x)\,dx = C_1$$

This is a linear first-order equation and can, therefore, be solved by the methods of section 2.3.

Just as in the case of first-order equations, we can sometimes find an integrating factor $\eta(x, y, y')$ for an equation of the form (4–52) which is not exact. Thus, a function $\eta(y', y, x)$ is an integrating factor of equation (4–52) if

$$\eta F = 0$$

is an exact equation.

For example, Liouville's equation

$$y'' + g(y) y'^2 + f(x) y' = 0$$

is not exact; but, upon multiplying through by $\eta = 1/y'$, we obtain the equation

$$\frac{1}{y'} y'' + g(y) y' + f(x) = 0 \qquad (4\text{--}61)$$

which can be written as

$$\frac{d}{dx} \left[\ln y' + Y(y) + X(x)\right] = 0$$

where $Y(y) = \int g(y)dy$ and $X(x) = \int f(x)dx$.

Equation (4–61) is therefore exact and its first integral is

$$\phi = \ln y' + Y + X = C_0$$

But, upon putting $C_1 = e^{C_0}$, we obtain the separable equation

$$y' e^Y = C_1 e^{-X}$$

which can be immediately integrated to obtain the general solution

$$\int e^{Y(y)} dy = C_1 \int e^{-X(x)} dx + C_2$$

to Liouville's equation.

Let $p_0(x)$, $p_1(x)$, and $p_2(x)$ be the coefficients of the linear homogeneous equation

$$p_0 y'' + p_1 y' + p_2 y = 0 \qquad (4\text{-}62)$$

and let $\eta(x)$ be a nontrivial solution of the equation

$$p_0 \eta'' + (2p_0' - p_1)\eta' + (p_0'' - p_1' + p_2)\eta = 0 \qquad (4\text{-}63)$$

which can also be written as

$$(\eta p_0)'' - (\eta p_1)' + \eta p_2 = 0 \qquad (4\text{-}64)$$

Since equation (4–60) is a necessary and sufficient condition that equation (4–3) be exact, equation (4–64) shows that the equation

$$(\eta p_0) y'' + (\eta p_1) y' + (\eta p_2) y = 0$$

is exact. But this implies that η is an integrating factor for equation (4–62). The linear homogeneous equation (4–63) for the integrating factor η is known as the *adjoint equation* of equation (4–62).

4.2 EQUATIONS WHICH ARE EQUIVALENT TO A LINEAR EQUATION

4.2.1 Equations Which Can Be Transformed Into a Third-Order Linear Equation

In section 2.4, we have seen that the first-order Riccati equation could be transformed into a second-order linear equation by the change of variable

$$u = e^{-\int y(x)h(x)dx} \qquad (4\text{-}65)$$

where $h(x)$ is not identically zero in the interval of interest.

Since linear equations are generally much easier to solve than nonlinear equations, the additional complexity incurred by raising the order of the equation is often justified. We shall now show that the change of variable (4–65) transforms the second-order nonlinear equation

$$y'' + p(x)y'' - 3h(x)yy' = f(x) + g(x)y + [h'(x) + p(x)h(x)]y^2 - h^2(x)y^3$$

$$(4\text{-}66)$$

where $p(x)$, $h(x)$, and $g(x)$ can be any functions of x, into a linear equation of the third order. To this end we differentiate equation (4–65) three times in succession to obtain

$$u' = -yh(x)u \tag{4–67}$$

$$u'' = (y^2h^2 - y'h + yh')u \tag{4–68}$$

$$u''' = -h\left(y'' + 3hyy' + \frac{2h'}{h}y' + \frac{h''}{h}y - 3y^2h' + h^2y^3\right)u \tag{4–69}$$

Substituting for y'' from equation (4–66) into equation (4–69) gives

$$u''' = -h\left[\left(2\frac{h'}{h} - p\right)y' + \left(\frac{h''}{h} + g\right)y + f + (ph - 2h')y^2\right]u$$

And using equation (4–68) to eliminate y' in this equation gives

$$u''' + \left(p - \frac{2h'}{h}\right)u'' + hfu = -hyu\left(\frac{h''}{h} + g - \frac{2h'^2}{h^2} + p\frac{h'}{h}\right)$$

Hence, it follows from equation (4–67) that u satisfies the third-order linear equation

$$u''' + p_1(x)u'' + q_1(x)u' + r_1(x)u = 0 \tag{4–70}$$

where

$$p_1(x) = p - 2\frac{h'}{h}$$

$$q_1(x) = \frac{h''}{h} + g - 2\frac{h'^2}{h^2} + p\frac{h'}{h}$$

$$r_1(x) = hf$$

For example, consider the equation

$$y'' + 3yy' + y^3 = 0 \qquad (4\text{-}71)$$

This equation is of the form (4-66), with $p = f = g = 0$ and $h = -1$. Hence, $p_1 = q_1 = r_1 = 0$, and equation (4-70) becomes

$$u''' = 0$$

But this equation has the solution

$$u = C_0 + C_1 x + C_2 x^2$$

which can be substituted into equation (4-67) to show that equation (4-71) has the general solution

$$y = \frac{2x + b_1}{x^2 + b_1 x + b_0}$$

where we have put $b_1 = C_1/C_2$ and $b_0 = C_0/C_2$.

4.2.2 Equations Which Are Equivalent to a Second-Order Linear Equation

We shall now consider a class of second-order nonlinear equations whose solutions can be expressed in terms of the solutions of second-order linear equations. The work on this problem began with Painlevé (ref. 11) and was carried on by Herbst (ref. 12), Gergen and Dressel (ref. 13), and Pinney (ref. 14). We shall present only the results here without proving any of the assertions. The references given should be consulted for details. Thus, for any functions $w(x)$ and $q(x)$ of x, any function $Y(y)$ of y, and any constant a, the differential equation

$$y'' = \frac{1 - Y'(y)}{Y(y)} y'^2 - \frac{w'(x)}{w(x)} y' = \left[q(x) + aw^2(x) \exp\left(-4 \int \frac{dy}{Y(y)} \right) \right] Y(y) \qquad (4\text{-}72)$$

has a solution of the form

$$y = \Theta(\omega^{1/2}) \qquad (4\text{-}73)$$

where Θ is the inverse of the function θ defined by

$$\theta(y) = \exp \int \frac{dy}{Y(y)} \tag{4-74}$$

(that is, $\Theta[\theta(y)] = y$) and the function ω is given by

$$\omega = C_1 u_1^2 + C_2 u_2^2 + C_3 u_1 u_2 \tag{4-75}$$

where C_1, C_2, and C_3 are constants and u_1 and u_2 are two linearly independent solutions of the linear differential equation

$$u'' - \frac{w'(x)}{w(x)} u' - q(x)u = 0 \tag{4-76}$$

For example, consider the equation introduced by Painlevé (ref. 11)

$$y'' + \frac{b}{y} y'^2 + f(x)y' = g(x)y \tag{4-77}$$

where b is a constant and f and g are any functions of x. This is a special case of equation (4–72) with $(1-Y')/Y = b/y$, $w'/w = -f$, $a = 0$, and $q = (1+b)g(x)$. Hence, we can take $Y = y/(b+1)$ and equation (4–74) becomes

$$\theta = \exp\left[(b+1) \int \frac{dy}{y} \right] = \exp\left[(b+1) \ln y \right] = y^{(b+1)}$$

Therefore,

$$\Theta(\theta) = \theta^{1/(b+1)}$$

In this case the differential equation (4–76) is

$$u'' + fu' - (1+b)gu = 0 \tag{4-78}$$

And the solution to equation (4–77) is given by

$$y = \Theta(\omega^{1/2}) = (\omega^{1/2})^{1/(1+b)} = (C_1 u_1^2 + C_2 u_2^2 + C_3 u_1 u_2)^{1/[2(1+b)]}$$

$$(4\text{--}79)$$

where u_1 and u_2 are two linearly independent solutions of equation (4–78). It can be verified by direct substitution and by applying equation (4–78) that equation (4–77) will be satisfied if, and only if,

$$C_3 = 2\sqrt{C_1}\,\sqrt{C_2}$$

Hence, equation (4–79) becomes

$$y = (\sqrt{C_1}\,u_1 + \sqrt{C_2}\,u_2)^{1/(1+b)}$$

But since $\sqrt{C_1}\,u_1 + \sqrt{C_2}\,u_2$ is also a solution of equation (4–78), this is equivalent to taking $y = u^{1/(1+b)}$ where u is a general solution of equation (4–78).

Next consider the equation (ref. 14)

$$y'' + p(x)y + My^{-3} = 0 \tag{4--80}$$

where $p(x)$ is any function of x and M is a constant. This is a special case of equation (4–72) with $w = \text{constant} = C_0$, $Y = y$, $q(x) = -p(x)$, $a = M/C_0^2$. Hence, $\theta(y) = y$ and, therefore, its inverse is $\Theta(\theta) = \theta$. But this shows that

$$y = \Theta\,(\omega^{1/2}) = \omega^{1/2} = (C_1 u_1^2 + C_2 u_2^2 + C_3 u_1 u_2)^{1/2} \tag{4--81}$$

where u_1 and u_2 are linearly independent solutions of

$$u'' + p(x)u = 0 \tag{4--82}$$

Let J denote the Wronskian of the solutions u_1 and u_2; that is,

$$J = u_1 u_2' - u_1' u_2 \tag{4--83}$$

Upon differentiating equation (4–83) with respect to x and using the fact that u_1 and u_2 satisfy equation (4–82), we find that [42]

---

[42] Or see section 5.9.

$$\frac{dJ}{dx}=0$$

which shows that $J=$ constant. Substituting the solution (4–81) into equation (4–80), using equation (4–82), and using the fact that its Wronskian is constant, we find that equation (4–81) is a solution of equation (4–80) if, and only if,

$$C_3^2=2\left(\frac{M}{J^2}+C_1C_2\right)$$

4.3 MISCELLANEOUS METHODS

4.3.1 Change of Variables

Frequently, a good choice of new dependent and independent variables will convert an equation to a simpler and more easily analyzed form. Some procedures which have proved helpful for this purpose are given in this section.

4.3.1.1 *General transformation of linear equations.* — The second-order linear equation (4–3) is transformed by introducing a new dependent variable v which is of the form

$$v=f(x)y \tag{4–84}$$

and a new independent variable t which is of the form

$$t=g(x) \tag{4–85}$$

into an equation which is also linear and of the second order.

The change of variable (4–84) transforms the linear homogeneous equation

$$y''+p(x)y'+q(x)y=0 \tag{4–36}$$

into an equation

$$v''+p_1(x)v'+q_1(x)v=0 \tag{4–86}$$

of the same type. It is clear that equation (4–86) can be transformed back into the original equation (4–36) by a change of variable which is also of the form (4–84). Any two equations which can be transformed into one another by a change of variable of the type (4–84) are said to be *equivalent*. The new coefficients $p_1(x)$ and $q_1(x)$ of equation (4–86) are related to the original coefficients $p(x)$ and $q(x)$ by

$$q - \frac{1}{2}p' - \frac{1}{4}p^2 = q_1 - \frac{1}{2}p_1' - \frac{1}{4}p_1^2$$

Hence, we may say that the quantity

$$\mathscr{I} \equiv q - \frac{1}{2}p' - \frac{1}{4}p^2 \tag{4–87}$$

remains invariant under the transformation (4–84). It is therefore called the *invariant* of equation (4–36). We have seen that any two equivalent equations have the same invariant. It can also be shown that, conversely, any two equations with the same invariant are equivalent.

If the solution of one equation of the form (4–36) is known, it is possible to find the solution of all equations which have the same invariant. Thus, when a new equation is encountered, we can compare its invariant with those of equations whose solutions are known. If we can find one which is the same, we will have succeeded in solving the original equation. In particular, the adjoint equation (see section 4.1.6) of equation (4–36) has the same invariant $\mathscr{I}$ as equation (4–36) and is therefore equivalent to it.

If the coefficient $p(x)$ in equation (4–36) is identically equal to zero, we say that the equation is in *normal*[43] *form*. In this case the invariant (4–87) is given by

$$\mathscr{I} = q$$

In the general case the equation (4–36) with invariant (4–87) can always be transformed into the normal equation

$$v'' + \mathscr{I}v = 0 \tag{4–88}$$

by a change of variable of the form (4–84). Then equation (4–88) is said to be the

---

[43]Notice that in this case the meaning of term *normal* is different from that of section 1.4. The proper interpretation of this term should always be clear from the context.

normal form of equation (4–36). It is clear that any given equation has only one normal form and that all equivalent equations have the same normal form.

The normal differential equation

$$\frac{d^2y}{dx^2} + \mathscr{I}(x)y = 0$$

can always be transformed into the normal equation

$$\frac{d^2v}{dt^2} + J(t)v = 0$$

by changing both the dependent and independent variables by transformations of the types (4–84) and (4–85), respectively, provided the nonlinear differential equation

$$\left(\frac{dt}{dx}\right)^4 J(t) = \left(\frac{dt}{dx}\right)^2 \mathscr{I}(x) - \frac{1}{2}\frac{d^3t}{dx^3}\frac{dt}{dx} + \frac{3}{4}\left(\frac{d^2t}{dx^2}\right)^2$$

can be solved for the new independent variable t. When this is the case, the function f in equation (4–84) which determines the new dependent variable is

$$f(x) = \sqrt{\frac{dt}{dx}}$$

A fuller discussion of this topic as well as proofs of the various assertions made in this section can be found in Rainville (ref. 15, chapter 1).

4.3.1.2 *Transformation to an equation with constant coefficients.*—We have seen that the homogeneous Euler equation (section 4.1.3.2) can be transformed by a change of independent variable into an equation with constant coefficients (which we know how to solve). More generally, the linear homogeneous equation

$$y'' + p(x)y' + q(x)y = 0 \tag{4–36}$$

can be transformed into a linear equation with constant coefficients by a change of independent variable if, and only if,

$$\frac{\dfrac{dq}{dx}+2pq}{q^{3/2}} = \text{constant} \tag{4-89}$$

When condition (4-89) holds, the change in variable (4-85) is given by

$$t = c \int \sqrt{q(x)}\,dx \tag{4-90}$$

where c is any constant.

In order to prove this, it is only necessary to substitute the change in variable (4-85) into equation (4-36) and obtain the transformed equation

$$\frac{d^2y}{dt^2} + \frac{\dfrac{d^2t}{dx^2}+p\,\dfrac{dt}{dx}}{\left(\dfrac{dt}{dx}\right)^2}\frac{dy}{dt} + \frac{g}{\left(\dfrac{dt}{dx}\right)^2}\,y = 0$$

It is now easy to see that the coefficient of y in this equation will be constant if, and only if, t is given by equation (4-90). Further, substituting (4-90) into the coefficient of dy/dt, we find that this coefficient will also be constant if, and only if, condition (4-89) is satisfied.

Thus, for example, for the homogeneous Euler equation, $p(x) = p_0/x$ and $g(x) = q_0/x^2$. Hence, equation (4-89) is satisfied, and equation (4-90) for the new independent variable becomes in this case

$$t = c \int \frac{q_0^{1/2}}{x}\,dx = cq_0^{1/2} \ln x$$

The change of variable (4-30) given in section 4.1.3.2 is a special case of this.

4.3.1.3 *Interchanging of dependent and independent variables.*—Differentiating the identity

$$\frac{dy}{dx} = \left(\frac{dx}{dy}\right)^{-1}$$

with respect to x shows that

$$\frac{d^2y}{dx^2} = -\frac{d^2x}{dy^2}\left(\frac{dx}{dy}\right)^{-3}$$

And when these relations are substituted into the general differential equation (4–1), we obtain the equation

$$F\left[-\frac{d^2x}{dy^2}\left(\frac{dx}{dy}\right)^{-3}, \left(\frac{dx}{dy}\right)^{-1}, y, x\right] = 0$$

in which x is the dependent variable and y is the independent variable. Sometimes this change of variable will result in an equation whose solution is known or can be found.

4.3.1.4 *Legendre transform.* — The *Legendre transform* which consists of introducing the new independent and dependent variables p and q, respectively, defined by

$$p \equiv \frac{dy}{dx} \tag{4–91}$$

$$q \equiv x\frac{dy}{dx} - y \tag{4–92}$$

can be used to radically alter the form of a differential equation. It follows from these relations, after differentiating equation (4–92), that

$$dq = xdp + pdx - dy \qquad\qquad dy = pdx$$

And when these equations are solved for x and y, we find that the inverse transformation is given by

$$x = \frac{dq}{dp} \tag{4–93}$$

$$y = p\frac{dq}{dp} - q \tag{4–94}$$

But it follows from equations (4–91) and (4–93) that

$$\frac{d^2y}{dx^2} = \left(\frac{dx}{dp}\right)^{-1} = \left(\frac{d^2q}{dp^2}\right)^{-1} \tag{4-95}$$

Hence, substituting equation (4–91) and equations (4–92) to (4–95) into equation (4–1) yields the transformed differential equation

$$F\left(\left(\frac{d^2q}{dp^2}\right)^{-1}, p, p\frac{dq}{dp} - q, \frac{dq}{dp}\right) = 0$$

which is also a second-order equation. If a solution $q = f(p)$ of this equation can be found, it can be substituted into equations (4–93) and (4–94) to obtain the parametric equations

$$y = pf'(p) - f(p)$$

$$\bar{x} = f'(p)$$

of a solution y of equation (4–1).

4.3.2 Equation Splitting

When equations which are split in some natural way into sums, quotients, or products of terms, such as

$$f(y'', y', y, x) + g(y'', y', y, x) = 0$$

or

$$\frac{f(y'', y', y, x)}{g(y'', y', y, x)} = c$$

are encountered, it is sometimes possible to obtain a solution by putting

$$f(y'', y', y, x) = h(x) = -g(y'', y', y, x)$$

for equations of the first type and

$$f(y'', y', y, x) = ch(x)$$

$$g(y'', y', y, x) = h(x)$$

for equations of the second type. If $h(x)$ can be chosen so that a common solution to the pair of equations can be found, this solution will also be a solution of the original equation.

4.3.3 **Tables of Differential Equations and Solutions**

Two valuable catalogs of solutions to differential equations can be found in the volumes by Murphy (ref. 16) and Kamke (ref. 17). Murphy lists over 2000 solved equations which are classified according to order and degree; Kamke gives about 1500 equations along with their solutions and references to the literature.

CHAPTER 5

Review of Complex Variables

A general procedure for solving second-order linear equations will be given in chapter 6. But this procedure involves the use of power series, and the discussion of power series becomes much simpler when it is carried out in its natural setting—the complex plane. In order to take advantage of this fact we shall extend the definition of a differential equation given in chapter 1 to include the case where the variables which occur in the equation are complex. Thus, by considering a more general situation we are actually able to simplify the treatment. Another reason for making this extension is that it allows us to see how solutions are connected across the singular points of the equation.

In this chapter, those concepts from the theory of functions of a complex variable which are needed for this purpose will be reviewed. The treatment is essentially descriptive; and rigorous proofs of the various assertions are, for the most part, omitted. For a more detailed treatment of the topics covered herein (including the omitted proofs), as well as a more complete coverage of the vast field of complex variables, the reader is referred to the many excellent texts [44] which are devoted entirely to this subject.

5.1 COMPLEX VARIABLES

Let x and y be two independent real variables and, as is the usual practice, put $i = \sqrt{-1}$. Then $z = x + iy$ is a complex variable. It is frequently convenient to think of the values of z as points in a plane, called the *complex plane*, whose Cartesian coordinates are x and y.

---

[44] A good elementary treatment is given by Churchill (ref. 18). A more advanced and theoretical treatment is given by Ahlfors (ref. 19), while the text by Carrier, Krook, and Pearson (ref. 20) emphasizes advanced applications.

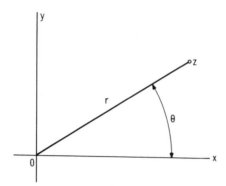

FIGURE 5-1.—Polar representation of complex number z.

Instead of using the Cartesian coordinates x and y, we can also use the polar coordinates $r=\sqrt{x^2+y^2}$ and $\theta=\tan^{-1} y/x$ to locate points in this plane. (The relation between the polar and Cartesian coordinates is illustrated in fig. 5-1.) Then

$$z=x+iy=r \cos \theta+ir \sin \theta$$

And upon using Euler's formula, we obtain the polar representation

$$z=r (\cos \theta+i \sin \theta) = re^{i\theta}$$

of the complex number z. Notice that for $n=0, \pm 1, \pm 2, \ldots$

$$e^{i2n\pi} = \cos 2n\pi+i \sin 2n\pi=1$$

Hence,

$$z=re^{i\theta}= re^{i\theta}e^{i2n\pi}= re^{i(\theta+2n\pi)}$$

The definitions of a domain and a neighborhood have been given in section 1.1 for a general n-dimensional space. We shall continue to use these definitions in the two-dimensional complex plane.

The complex conjugate of the variable z is denoted by z^* and defined by $z^* = x - iy$.

The *absolute value* or *modulus* of the complex variable z is defined to be the length of the vector joining the origin with the point z in the complex plane and is denoted by $|z|$. Therefore,[45]

$$|z| = \sqrt{x^2 + y^2} = \sqrt{zz^*}$$

5.2 ANALYTIC FUNCTIONS OF COMPLEX VARIABLE

Let $u(x, y)$ and $v(x, y)$ be any two real-valued functions[46] (of the variables x and y) which are defined in some region of the complex plane. Then $w = u + iv$ is a complex-valued function of x and y. Since w associates a complex number with each point $z = x + iy$ of some region of the complex plane, we say that w *is a function of the complex variable* z.

We shall consider only a particular class of functions of a complex variable called *analytic* or *holomorphic* functions. In order to define this class of functions we first introduce the concept of a complex derivative. To this end, let $w = u + iv$ be a function of the complex variable z and suppose that x and y are changed by the amounts Δx and Δy, respectively. Then w changes by an amount $\Delta w = \Delta u + i\,\Delta v$. Now, by analogy with the definitions of the derivative of a real-valued function of a real variable, we define dw/dz, *the derivative*[47] *of* w *with respect to* z at the point z, to be the limit

$$\frac{dw}{dz} = \lim_{\substack{\Delta x \to 0 \\ \Delta y \to 0}} \frac{\Delta u + i\,\Delta v}{\Delta x + i\,\Delta y} = \lim_{\substack{\Delta x \to 0 \\ \Delta y \to 0}} \frac{\Delta w}{\Delta z}$$

provided that this limit not only exists but that it is independent of the manner in which Δx and Δy approach zero. When Δx and Δy approach zero in some prescribed manner, Δz approaches zero along some path in the complex plane,

---

[45] Notice that $|z|$ is equal to the polar coordinate r.

[46] Recall that according to the convention adopted in section 1.1 we assume that all functions are single valued unless explicitly stated otherwise.

[47] We shall frequently write $w'(z)$ or w' in place of dw/dz and $u^{(n)}(z)$ or $u^{(n)}$ in place of $d^n w/dz^n$ for $n = 1, 2, \ldots$

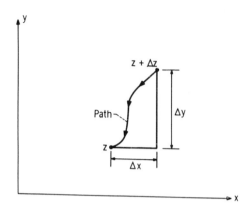

FIGURE 5–2.—Typical path along which Δz can approach zero.

as indicated schematically in figure 5–2. But the definition implies that, if w is to have a derivative at the point z, then $\Delta w/\Delta z$ must approach the same limit for every such path along which Δz approaches zero. Although this requirement imposed on the complex derivative may seem unimportant, its implications are enormous. In fact, by allowing Δz to approach zero along various paths, it can be readily shown (ref. 18, p. 34) that, if u and v are continuously differentiable,[48] a necessary and sufficient condition for the existence of the derivative dw/dz at the point z_0 is that u and v satisfy the two *Cauchy-Riemann equations*

$$\frac{\partial u}{\partial x} = \frac{\partial v}{\partial y} \quad \text{and} \quad \frac{\partial u}{\partial y} = -\frac{\partial v}{\partial x}$$

at this point. This shows that much greater restrictions are imposed on those complex functions which possess complex derivatives in the sense of the definition given above than are imposed on the real functions which possess ordinary real derivatives. However, since the complex derivative is formally the same as the real derivative, the usual rules for differentiating sums, products, quotients, etc., still apply (ref. 18, p. 31).

A function $w(z)$ of the complex variable z is said to be *analytic*[49] or *holomorphic in a domain* D if it possesses a derivative at every point of *D*. Frequently,

---

[48] This means that the partial derivatives $\partial u/\partial x$, $\partial u/\partial y$, $\partial v/\partial x$, and $\partial v/\partial y$ exist and are continuous.

[49] The terms *regular* and *monogenic* are also used.

when it is of no consequence in the discussion, the reference to the domain D is omitted and we simply say that the function $w(z)$ is analytic. A function is said to be *analytic at a point* z_0 if it is analytic in some neighborhood of this point.

Notice that the identity function

$$w = z = x + iy$$

is analytic at every point since the Cauchy-Riemann equations (with $u = x$ and $v = y$) are always satisfied. However, the complex conjugate of this function

$$w = z^* = x - iy$$

is not analytic at any point since in this case $\partial u / \partial x = 1$ and $\partial v / \partial y = -1$ and therefore the Cauchy-Riemann equations are never satisfied.

The real and imaginary parts of the function

$$w = \frac{1}{z} = \frac{1}{x + iy} = \frac{z^*}{zz^*} = \frac{x - iy}{x^2 + y^2}$$

are

$$u = \frac{x}{x^2 + y^2} \qquad \text{and} \qquad v = \frac{-y}{x^2 + y^2}$$

respectively. And upon taking the partial derivatives of these functions we see that the Cauchy-Riemann equations are satisfied at every point except $z = 0$, where the partial derivatives fail to exist. Hence, $w = 1/z$ is analytic at every point except $z = 0$.

It is a remarkable fact (see ref. 18, p. 122) that any function $w(z)$ which is analytic at a point z_0 possesses derivatives of all orders at this point. And these derivatives are themselves analytic functions at this point. It is easy to see that the sum and product of any two analytic functions are analytic within any domain in which both functions are analytic. Usually, any function which is obtained from a real algebraic, elementary-transcendental function (trigonometric, exponential, logarithmic, etc.) or a common higher-transcendental function (Bessel function, hypergeometric function, etc.) by replacing the real variable x by the complex variable $z = x + iy$ is an analytic function within some domain.

5.3 CONFORMAL MAPPING [50]

According to the definition given in section 5.2, a complex function $w = u + iv$ of the complex variable $z = x + iy$ associates a pair of numbers (u, v) with each point (x, y) of some region of the complex z-plane. We can interpret these numbers as coordinates of a point in a complex w-plane. Thus, we may think of the function $w(z)$ as a transformation or a mapping of some region in the z-plane into some region in the w-plane. More specifically, we can think of an analytic function $w(z)$ which is defined on a domain D in the z-plane as a mapping of D onto a region R in the w-plane, as shown schematically in figure 5–3. In order to determine the properties of an analytic function $w(z)$ it is frequently helpful to study the manner in which this function transforms various points, curves, or domains in the z-plane into corresponding points, curves, or regions in the w-plane.

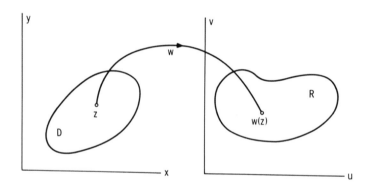

FIGURE 5–3. — Mapping of D onto R by $w(z)$.

Thus, the transformation $w = 1/z$ associates a single point in the w-plane with each point in the z-plane except the origin $z = 0$. It maps all those points lying outside the circle with radius R and center at the origin in the z-plane into the interior of the circle with radius $1/R$ and center at the origin in the w-plane. Although there is no point of the z-plane which maps into the point $w = 0$, we can, by choosing R sufficiently large, make the points outside this

---

[50] A comprehensive treatment of this subject can be found in Nehari (ref. 21).

circle map into a circle of arbitrary small radius about $w=0$. However, it is frequently convenient to "complete" the z-plane by adding a fictitious point $z=\infty$ which maps into the point $w=0$ under the mapping $w=1/z$. Thus, when we speak of the behavior of an equation or a function at the point $z=\infty$, we actually mean the behavior at the point $w=0$ of the transformed equation or function obtained by putting $z=1/w$. When it is necessary to distinguish between the point $z=\infty$ and the other points of the complex plane, the latter are said to be *finite* points. When the complex plane includes the point $z=\infty$, it is called the *extended plane*; and when $z=\infty$ is excluded, it is called the *finite plane*.

Let $w(z)$ be analytic in a domain D and let z_0 be any point of D at which $dw/dz \neq 0$. Then $w(z)$ transforms any two smooth curves passing through z_0 in such a way that their image curves intersect at the point $w_0=w(z_0)$ with the same angle (in both magnitude and sense of rotation) as the original curves in the z-plane (ref. 18, p. 174). Thus, the mapping "preserves angles," and we say that $w(z)$ is a *conformal mapping* at all points where $dw/dz \neq 0$.

A mapping which is particularly useful for the treatment of certain types of linear differential equations is the *linear fractional transform*

$$w = \frac{az+b}{cz+d} \tag{5–1}$$

Notice that, if $ad-bc=0$, this transformation reduces to $w=$ constant. In any other case, it transforms each point in the extended z-plane into a point in the extended w-plane in such a way that no two points in the z-plane map into the same point in the w-plane. In addition, there is a point in the z-plane which maps into each point in the w-plane. For this reason, we say that a linear fractional transformation with $ad-bc \neq 0$ is *nonsingular*.

If we consider straight lines and points as being degenerate circles (i.e., circles having infinite or zero radii), we can say that the *linear fractional transformation always maps circles into circles.*

Performing two nonsingular linear fractional transformations in succession is equivalent to performing a single nonsingular linear fractional transformation. In fact, any nonsingular linear fractional transformation can be performed by carrying out not more than four successive transformations each of which has one of the three elementary forms

$$w = \alpha z \qquad \text{(rotation and stretching)} \tag{5–2}$$

117

DIFFERENTIAL EQUATIONS

$$w = \beta + z \quad \text{(translation)} \tag{5-3}$$

$$w = \frac{1}{z} \quad \text{(inversion)} \tag{5-4}$$

In order to prove this, first suppose that $c \neq 0$. Then equation (5-1) can be written as

$$w = \frac{a}{c} + \frac{\left(\dfrac{b}{c}\right) - \left(\dfrac{ad}{c^2}\right)}{z + \left(\dfrac{d}{c}\right)} \tag{5-5}$$

Now transform the z-plane into the t_1-plane by the transformation

$$t_1 = z + \left(\frac{d}{c}\right)$$

Then transform the t_1-plane into the t_2-plane by

$$t_2 = \frac{1}{t_1}$$

and the t_2-plane into the t_3-plane by

$$t_3 = \left(\frac{b}{c} - \frac{ad}{c^2}\right) t_2$$

Finally, transform the t_3-plane into the w-plane by

$$w = \frac{a}{c} + t_3$$

Upon combining these successive transformations we obtain equation (5-5) and therefore equation (5-1). This proves the assertion for the case where $c \neq 0$. When $c = 0$, equation (5-1) reduces to

$$w = \frac{a}{d}z + \frac{b}{d}$$

which is easily seen to be equivalent to the succession of transforms

$$t_1 = \frac{a}{d}z \quad \text{and} \quad w = t_1 + \frac{b}{d}$$

If the transformation (5–1) is to be nonsingular, at least two of the constants a, b, c, and d must be nonzero. By dividing through by one of these, it is easy to see that equation (5–1) in fact contains only three arbitrary constants. It is therefore not surprising that these constants can always be chosen so that the linear fractional transformation maps *any* three given points in the z-plane, say z_1, z_2, and z_3, into any three given points, say w_1, w_2, and w_3, in the w-plane. Thus, for example, the linear fractional transformation which takes $z = z_1$ into $w = 0$, $z = z_2$ into $w = 1$, and $z = z_3$ into $w = \infty$ is

$$w = \frac{z_2 - z_3}{z_2 - z_1} \frac{z - z_1}{z - z_3}$$

5.4 ISOLATED SINGULAR POINTS OF ANALYTIC FUNCTIONS

If a function $w(z)$ is analytic at every point in some neighborhood of a point z_0 except at the point z_0 itself,[51] then z_0 is called an *isolated singular point* or an *isolated singularity* of the function $w(z)$. Thus, the function

$$w(z) = \frac{z + 2}{z(z + 1)^2} \tag{5–6}$$

has isolated singularities at the points $z = 0$ and $z = -1$.

An isolated singular point z_0 of the function $w(z)$ is called a *removable singularity* if the limit of $w(z)$ as $z \to z_0$ is equal to some finite number. Let k be a positive integer. An isolated singular point z_0 of the function $w(z)$ is said to be *a pole of order k* if the limit as $z \to z_0$ of the quantity $(z - z_0)^k w(z)$ is equal to some finite nonzero number. A pole of order one is called a *simple pole*.

---

[51] The function need not even be defined at the point z_0.

Thus, for example, the function given by equation (5–6) has a simple pole at $z = 0$ and has a pole of order two at $z = -1$.

Because of the way analytic functions arise in practice, it turns out that we sometimes arrive at a function $w(z)$ which is not defined at a point z_0 but is defined and analytic at every point of a neighborhood of z_0. The point z_0 is thus an isolated singular point of $w(z)$. But if this point is also a removable singularity, the function can be made analytic at z_0 simply by assigning a suitable value to $w(z)$ at this point (ref. 18, p. 158). For example, since division by zero is undefined, the function $w(z)$ given in equation (5–6) is undefined at $z = 0$. Thus, $0 \times w(0)$ is also undefined. Hence, the function $\zeta(z)$ defined by

$$\zeta(z) \equiv zw(z)$$

is not defined at the point $z = 0$. However, upon defining $\zeta(0)$ by

$$\zeta(0) \equiv \lim_{z \to 0} \zeta(z) = 2$$

we obtain a function which is analytic at $z = 0$.

It is easy to see that, if $w(z)$ has a pole of order k at z_0, the function

$$\zeta(z) \equiv (z - z_0)^k w(z)$$

has a removable singularity at z_0 and is therefore "essentially" analytic at this point.[52] We shall sometimes say that an analytic function has a pole of order zero at the point z_0 if z_0 is a removable singularity of this function.

Any isolated singular point of an analytic function which is not a pole or a removable singularity is called an *essential singularity*. For example, the function $\sin(1/z)$ has an essential singularity at $z = 0$.

There are some important differences between poles and essential singularities. For example, if $w(z)$ has a pole at z_0, the function $1/w(z)$ is analytic [53]

---

[52] For the purposes of this book we can assume that any analytic function which is encountered has already been defined at its removable singularities in such a way that it is analytic at these points.

[53] In fact, it is equal to zero at z_0 (see preceding footnote).

at z_0; but if $w(z)$ has an essential singularity at z_0, so does $1/w(z)$ (ref. 22, p. 110). A pole, then, is a point where a function $w(z)$ is not analytic only because its modulus $|w(z)|$ becomes infinitely large at this point and for no other reason.

A function which is analytic at every *finite* point of the complex plane is called an *entire function*. And *Liouville's theorem* (ref. 18, p. 125) states that any entire function which is also analytic at infinity (see section 5.3) must, in fact, be equal to a constant.

A *polynomial* is a function of the form $a_0 + a_1 z + \ldots + a_n z^n$ where $a_0, \ldots, a_n$ are complex constants. It is an entire function, and it has a pole of order n at infinity. A *rational* function is the ratio of two polynomials (which may be chosen to have no linear factors in common). It is therefore a function of the form

$$w(z) = \frac{a_0 + a_1 z + a_2 z^2 + \ldots + a_n z^n}{b_0 + b_1 z + b_2 z^2 + \ldots + b_m z^m} \qquad a_n \neq 0; b_m \neq 0 \qquad (5\text{-}7)$$

and is analytic everywhere in the finite plane except at those points where its denominator is equal to zero. These points are poles of $w(z)$. If $m \geq n$, then $w(z)$ is analytic at the point $z = \infty$; otherwise it has a pole of order $n - m$ at this point.

A function which is analytic at every point of a domain D except at those points of D where it has poles is said to be *meromorphic* in D. For example, the rational function (5-7) is meromorphic in the entire finite plane. In fact, any function which is meromorphic in the entire finite plane and has a pole at infinity is necessarily a rational function (ref. 20, p. 60). A function which is meromorphic in the entire finite plane can have at most a finite number of poles in any domain D of finite size. However, it may have infinitely many poles if D is infinitely large.[54]

In view of Euler's formula (see section 5.1) it is natural to define the function e^z by the formula

$$e^z = e^{x+iy} = e^x \cos y + i e^x \sin y$$

Then e^z is an entire function and has an essential singularity at $z = \infty$. We can now define the functions $\sin z$ and $\cos z$ by the formulas

---

[54] For example, D could be the entire plane or the upper half plane.

DIFFERENTIAL EQUATIONS

$$\sin z = \frac{e^{iz}-e^{-iz}}{2i} \qquad \text{and} \qquad \cos z = \frac{e^{iz}+e^{-iz}}{2}$$

which extend in a natural way the definitions given to these functions when the variables are real. They are also entire functions with essential singularities at $z=\infty$. On the other hand, the function $1/\sin z$ is meromorphic in the entire finite plane; and its poles are located at the points $z=n\pi$ for $n=0,\pm1,\pm2$, etc. It also has an essential singularity at $z=\infty$.

5.5 POWER SERIES

A *power series about a point* z_0 is an infinite series of the form

$$\sum_{n=0}^{\infty} a_n(z-z_0)^n \tag{5-8}$$

in which the coefficients a_n can be any complex numbers. This series certainly converges at the point z_0, which may be the only point at which it actually does converge. Or the opposite extreme could occur and the series might converge at every point of the finite plane. In all other cases the series will converge at every point within a circle of radius R and center at z_0, called the *circle of convergence* of the series, and will diverge at every point which lies outside this circle. Thus (ref. 19), there exists a number R lying in the range $0 \le R \le \infty$ called the *radius of convergence* of the series such that the series (5-8) converges at all points z which satisfy the inequality [55] $|z-z_0| < R$ and diverges at all points z which satisfy the inequality [56] $|z-z_0| > R$. The question of whether the series is convergent for the points which satisfy the equality $|z-z_0|=R$ (i.e., points on the circle of convergence) is more subtle but unimportant for our purposes.

A power series with a nonzero radius of convergence R converges to an analytic function and can be differentiated term by term (i.e., the order of summation and differentiation can be interchanged) *at every point within its circle of convergence.* Thus, there exists a function $w(z)$ which is analytic at every point within the circle $|z-z_0| < R$ such that

---

[55] The series is, in fact, absolutely convergent at these points. This means that the series still converges when each of its terms is replaced by its absolute value. See ref. 19 for more details.
[56] This result was first established by Abel.

122

$$
\left.
\begin{aligned}
w(z) &= \sum_{n=0}^{\infty} a_n (z-z_0)^n \\
\frac{dw(z)}{dz} &= \sum_{n=1}^{\infty} n a_n (z-z_0)^{n-1}
\end{aligned}
\right\} \qquad \text{for } |z-z_0| < R
$$

It can also be shown (ref. 19) that, when the series (5–8) has a nonzero radius of convergence R, there exists a positive constant $M \neq \infty$ such that for any number $0 < r < R$ the coefficients of the series satisfy the *Cauchy estimates*

$$
|a_n| < \frac{M}{r^n} \qquad \text{for } n=0, 1, 2, \ldots
$$

This result will be used in the discussion of the solutions of differential equations in the next chapter.

Now *suppose that* w(z) *is analytic in the domain* D *and that* z_0 *is any point of* D. *Then the power series*

$$
\sum_{n=0}^{\infty} \frac{1}{n!} w^{(n)}(z_0)(z-z_0)^n \tag{5–9}
$$

(where $w^{(0)}(z_0) \equiv w(z)$*) converges to* $w(z)$ *at every point* z *within the largest circle centered at* z_0 *lying entirely within* D (ref. 18, p. 129). *This series is called a *Taylor series* expansion of $w(z)$ about z_0. Its radius of convergence is at least equal to the shortest distance between z_0 and the boundary of D. It may be larger than this but we have no guarantee that the series will converge to $w(z)$ at points which lie outside of D. The series representation (5–9) is unique in the sense that if $\sum_{n=0}^{\infty} b_n(z-z_0)^n$ is any power series which converges to $w(z)$ within any circle about z_0, then necessarily (ref. 20, p. 49)

$$
b_n = \frac{1}{n!} w^{(n)}(z_0) \qquad \text{for } n=0, 1, 2, \ldots
$$

Next, suppose that $w(z)$ is analytic at every point of a domain D except for a certain number of isolated singular points and let z_0 be a point of D at which $w(z)$ is analytic. Then the circle of convergence of the Taylor series

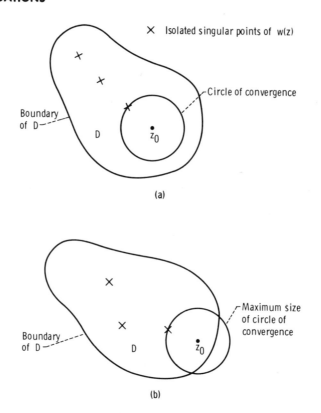

(a) Circle of convergence does not intersect boundary of D.
(b) Maximum circle of convergence intersects boundary of D.

FIGURE 5–4. — Circle of convergence of a Taylor series.

of $w(z)$ about z_0 passes through the nearest isolated singular point of $w(z)$ if it does not intersect the boundary of D. In this latter case we can only assert that the radius of convergence does not exceed the distance between z_0 and the nearest of the isolated singularities of $w(z)$ within D. These results are illustrated in figure 5–4. Thus, in particular, if the function $w(z)$ is analytic at every finite point of the complex plane except at a certain number of isolated singular points, then the circle of convergence of its Taylor series expansion about any point z_0 where $w(z)$ is analytic always passes through the nearest isolated singular point to z_0.

For example, we have seen in section 5.2 that $w(z) = e^z$ is an entire function. It is easy to see that the nth derivative of this function is

$$w^{(n)}(z) = e^z \qquad \text{for } n = 0, 1, 2, \ldots$$

Hence, the Taylor series expansion of e^z about $z = 0$ is

$$e^z = \sum_{n=0}^{\infty} \frac{1}{n!} z^n \tag{5-10}$$

And this series converges in the entire finite plane.

We have also shown in section 5.2 that the function $w(z) = 1/z$ is analytic at every point of the complex plane except the origin $z = 0$. The nth derivative of this function is

$$w^{(n)}(z) = \frac{(-1)^n n!}{z^{n+1}}$$

Hence, its Taylor series expansion about the point $z = 1$ is

$$\frac{1}{z} = \sum_{n=0}^{\infty} (-1)^n (z-1)^n$$

It is easy to verify that the radius of convergence of this series is equal to 1 and that the circle of convergence passes through the isolated singular point $z = 0$. Upon replacing $1 - z$ by z in this series we obtain the *geometric series*

$$\frac{1}{1-z} = \sum_{n=0}^{\infty} z^n \tag{5-11}$$

which converges within the unit circle $|z| = 1$. This circle passes through the isolated singular point $z = 1$ of the function $(1-z)^{-1}$.

We shall frequently find it necessary to add and multiply two power series. The sum of two power series about the same point can be obtained by adding the two series term by term. The resulting series will converge within the smaller of the two circles of convergence of the original series (ref. 23, p. 123). Now let $\sum_{n=0}^{\infty} c_n$ and $\sum_{n=0}^{\infty} d_n$ be any two absolutely convergent series. Any expression for the product $\left(\sum_{n=0}^{\infty} c_n \right) \left(\sum_{n=0}^{\infty} d_n \right)$ must certainly include

125

all terms of the type $c_i d_j$. But all terms of this type must belong to the array

$$c_0 d_0 + c_0 d_1 + c_0 d_2 + \ldots + c_0 d_n + \ldots$$

$$+ c_1 d_0 + c_1 d_1 + d_1 d_2 + \ldots + c_1 d_n + \ldots$$

$$+ c_2 d_0 + c_2 d_1 + c_2 d_2 + \ldots + c_2 d_n + \ldots$$

$$\vdots$$

$$+ c_m d_0 + c_m d_1 + c_m d_2 + \ldots + c_m d_n + \ldots$$

$$+ \ldots$$

And the series which is formed by grouping together the terms along the diagonals of this array and summing the result over all diagonals is called the *Cauchy product* of the two series. Thus, the general term in the Cauchy product is

$$a_n = c_n d_0 + c_{n-1} d_1 + c_{n-2} d_2 + \ldots + c_0 d_n = \sum_{k=0}^{n} c_{n-k} d_k = \sum_{k=0}^{n} c_k d_{n-k}$$

and the Cauchy product is the series $\sum_{n=0}^{\infty} a_n$. Cauchy's theorem (ref. 18, p. 147) states that *the Cauchy product $\sum_{n=0}^{\infty} a_n$ is absolutely convergent and that*

$$\left(\sum_{n=0}^{\infty} c_n \right) \left(\sum_{n=0}^{\infty} d_n \right) = \left(\sum_{n=0}^{\infty} a_n \right) = \sum_{n=0}^{\infty} \sum_{k=0}^{n} c_k d_{n-k}$$

The Cauchy product is particularly convenient to use when multiplying power series. For, in this case, we get

$$\left[\sum_{n=0}^{\infty} a_n (z - z_0)^n \right] \left[\sum_{n=0}^{\infty} b_n (z - z_0)^n \right] = \sum_{n=0}^{\infty} \sum_{k=0}^{n} a_k (z - z_0)^k b_{n-k} (z - z_0)^{n-k}$$

$$= \sum_{n=0}^{\infty} (z - z_0)^n \left(\sum_{k=0}^{n} a_k b_{n-k} \right)$$

which converges within the smaller circle of convergence of the two original series.

For example, let us use these ideas to find the power series expansion of the function $1/[(a-z)(b-z)]$, where a and b are arbitrary complex constants. To this end notice that the geometric expansion (5–11) implies that

$$\frac{1}{1-\dfrac{z}{a}} = \sum_{n=0}^{\infty} a^{-n}z^n \quad \text{and} \quad \frac{1}{1-\dfrac{z}{b}} = \sum_{n=0}^{\infty} b^{-n}z^n$$

Hence, upon forming the Cauchy product we obtain

$$\frac{1}{(a-z)(b-z)} = \frac{1}{ab} \sum_{n=0}^{\infty} z^n \sum_{k=0}^{n} \left(\frac{1}{a}\right)^k \left(\frac{1}{b}\right)^{n-k}$$

But since

$$\sum_{k=0}^{n} \left(\frac{1}{a}\right)^k \left(\frac{1}{b}\right)^{n-k} = \frac{\left(\dfrac{1}{a}\right)^{n+1} - \left(\dfrac{1}{b}\right)^{n+1}}{\dfrac{1}{a} - \dfrac{1}{b}}$$

we find that

$$\frac{1}{(a-z)(b-z)} = \frac{1}{b-a} \sum_{n=0}^{\infty} \left[a^{-(n+1)} - b^{-(n+1)}\right]z^n$$

Next, in order to find the Taylor series expansion of the function $\cos z/(1+z^2)$ about the point $z=0$, notice that by changing variables in the geometric expansion (5–11), we obtain the expansion

$$\frac{1}{1+z^2} = \sum_{n=0}^{\infty} (-1)^n z^{2n}$$

and that by using the expansion (5–9) and the definition of $\cos z$ given in section 5.4, we obtain the expansion

$$\cos z = \sum_{n=0}^{\infty} (-1)^n \frac{z^{2n}}{(2n)!}$$

Then, upon taking the Cauchy product of these two series, we obtain

$$\frac{\cos z}{1+z^2} = \sum_{n=0}^{\infty} (-1)^n z^{2n} \sum_{k=0}^{n} \frac{1}{(2k)!}$$

We have seen that a power series (with positive exponents) represents (converges to) an analytic function within a circle. Similarly, the series (ref. 18, p. 134)

$$\sum_{n=-\infty}^{\infty} a_n (z-z_0)^n \tag{5-12}$$

containing both positive and negative exponents, converges to an analytic function in an annular region[57] lying between two concentric circles centered at z_0 and of radii R_1 and R_2 with $R_1 < R_2$ (i.e., at all points z for which $R_1 < |z-z_0| < R_2$). Conversely, any function $w(z)$ which is analytic in an annular region $R_1 < |z - z_0| < R_2$ can always be represented by a series of the form (5-12) at every point of this region. This expansion is called a *Laurent series*. An important special case occurs when the function $w(z)$ has an isolated singularity at the point z_0 and is analytic at every other point within the circle $|z - z_0| = R$. In this case the series

$$w(z) = \sum_{n=-\infty}^{\infty} a_n (z-z_0)^n$$

converges to $w(z)$ at every point z of the punctured circular region $0 < |z-z_0| < R$ bounded by the circle $|z-z_0| = R$ and the point z_0. Then the isolated singularity z_0 *is a pole of order* k *of the function* w(z) *if, and only if,* $a_{-k} \neq 0$ *and* $a_n = 0$ *for all* n < − k. Hence, if z_0 is an essential singularity, infinitely many negative powers of $z - z_0$ will occur in the series.

For example, we can use the geometric series (5-11) to obtain the expansion

---

[57] Provided it converges at all.

$$\frac{1}{z-1} = \frac{1}{z}\frac{1}{1-\frac{1}{z}} = \sum_{n=0}^{\infty} z^{-n-1} = \sum_{n=1}^{\infty} z^{-n} = \sum_{n=-\infty}^{-1} z^{n}$$

of the function $(z-1)^{-1}$ about the point $z = \infty$. This series converges at all points which lie outside the circle $|z| = 1$. And since the series

$$\frac{3}{3-z} = \sum_{n=0}^{\infty} 3^{-n}z^{n}$$

converges within the circle $|z| = 3$, we conclude that the series

$$\frac{2z}{(z-1)(3-z)} = \frac{1}{z-1} + \frac{3}{3-z} = \sum_{n=-\infty}^{\infty} a_{n}z^{n}$$

with

$$a_{n} = \begin{cases} 1 & \text{for } n = -1, -2, \ldots \\ 3^{-n} & \text{for } n = 0, 1, 2, \ldots \end{cases}$$

converges in the annular region $1 < |z| < 3$.

5.6 COMPLEX INTEGRATION

Let $w(z)$ be analytic in a domain D. Then the integral $\int_{\Gamma} w(z)\,dz$ of $w(z)$ along a curve or path Γ which lies entirely within D is defined in terms of two real line integrals along Γ by

$$\int_{\Gamma} w(z)\,dz = \oint_{\Gamma} (u\,dx - v\,dy) + i\oint_{\Gamma} (u\,dy + v\,dx) \tag{5-13}$$

It is easy to verify from this definition that the usual rules of integration still apply. Thus, in particular, if α and β are complex constants and $w_{1}(z)$ and $w_{2}(z)$ are analytic functions

$$\int_\Gamma [\alpha w_1(z) + \beta w_2(z)]dz = \alpha \int_\Gamma w_1(z)dz + \beta \int_\Gamma w_2(z)dz$$

and if the direction of integration along Γ is reversed, the integral is multiplied by -1.

For example, the integral of the function $w(z) = z$ along the line $y_0 = $ constant, from the point $(0, y_0)$ to the point (x_0, y_0), is

$$\int_\Gamma z\,dz = \int_0^{x_0} x\,dx + i \int_0^{x_0} y_0\,dx = \frac{x_0^2}{2} + iy_0 x_0 = \frac{1}{2}(x_0 + iy_0)^2 - \frac{1}{2}(iy_0)^2$$

In order to integrate $w(z) = z$ along a circular path centered at $z = 0$, it is best to use polar notation. Thus, let the radius of the circle be R. Then on this circle $z = Re^{i\theta}$ and $dz = iRe^{i\theta}\,d\theta$. Hence,

$$\int_{r=R} z\,dz = i\int_0^{2\pi} R^2 e^{2i\theta}\,d\theta = 0$$

More generally, it can be shown (ref. 18, p. 111) that, if C is any closed curve within D and if $w(z)$ is analytic at every point in the interior of C, then

$$\int_C w(z)\,dz = 0$$

Therefore, if Γ_1 and Γ_2 are any two curves in D which begin at the point z_1 and end at the point z_2 (fig. 5–5), then

$$\int_{\Gamma_1} w(z)\,dz = \int_{\Gamma_2} w(z)\,dz$$

provided $w(z)$ is analytic within the domain enclosed by these curves.

Because of this fact the exact path along which the integration is carried out is frequently unimportant; and when this is the case, we write $\int_{z_1}^{z_2} w(z)\,dz$ in place of $\int_\Gamma w(z)\,dz$.

The function $w(z) = 1/z$ has a singular point at the origin; and its integral along the circle of radius R centered at the origin is

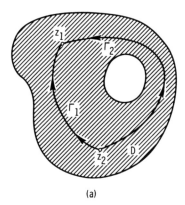
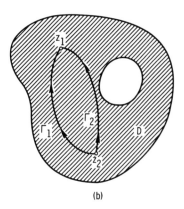

(a) Line integrals along Γ_1 and Γ_2 not necessarily equal.

(b) Line integrals equal along Γ_1 and Γ_2.

FIGURE 5–5.—Paths for line integral. (Arrows indicate direction of integration.)

$$\int_{r=R} \frac{1}{z}\, dz = i \int_0^{2\pi} d\theta = 2\pi i$$

If the function $w(z)$ is an analytic function of z in D, the function $F(z)$ defined by

$$F(z) = \int_{z_0}^z w(z)\, dz$$

is also an analytic function of z in D and (ref. 18, p. 114)

$$\frac{dF(z)}{dz} = w(z)$$

Hence,

$$F(z) - F(z_0) = \int_{z_0}^z F'(z)\, dz$$

This shows that just as in the case of real variables, integration and differentiation are inverse processes. For example, in order to evaluate the integral $\int_{z_0}^z z^n dz$ notice that, for $n \neq -1$,

$$z^n = \frac{1}{n+1} \frac{dz^{n+1}}{dz}$$

Hence,

$$\int_{z_0}^{z} z^n dz = \frac{1}{n+1} \int_{z_0}^{z} \frac{dz^{n+1}}{dz} \, dz = \frac{z^{n+1}}{n+1} - \frac{z_0^{n+1}}{n+1}$$

It can also be shown (ref. 18, p. 141) that any convergent power series can be integrated term by term and the resulting series will have the same circle of convergence as the original series. In fact, it can be shown that the product of any convergent power series about a point z_0 with a function of the form $(z - z_0)^\lambda$, where λ is a complex constant, can also be integrated term by term.

5.7 ANALYTIC CONTINUATION

5.7.1 Definition

First, suppose that $w_1(z)$ and $w_2(z)$ are both analytic in some common domain D. It can be shown (ref. 18, p. 259) that *if* $w_1(z) = w_2(z)$ *at all points of some subdomain of* D *or even at all points of some curve which lies entirely within* D, *then* $w_1(z)$ *and* $w_2(z)$ *are equal at* **every** *point of* D. This assertion is known as the *fundamental theorem of analytic continuation*. It means that *there is only one analytic function in a domain* D *which takes on any given set of values which are prescribed at every point of a subdomain of* D *or even at every point of some curve in* D. For example, if the function $w(z)$ is analytic in a domain D and is equal to zero at every point of a subdomain of D or at every point of a curve lying within D, then $w(z)$ is zero at every point of D.

We have seen in section 5.5 that the analytic function

$$\frac{1}{1-z} \tag{5-14}$$

which is defined and analytic at every point of the complex plane except $z = 1$ can be expanded in the Taylor series

$$\sum_{n=0}^{\infty} z^n \tag{5-15}$$

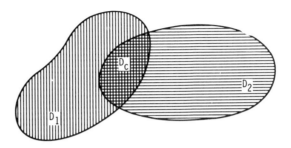

FIGURE 5–6. – Domains for direct analytic contination; D_c is the common part of both domains D_1 and D_2.

which converges only within the unit circle. Thus, the function (5–15) is defined only within the unit circle, where it is equal to the function (5–14), which, however, is defined in a much larger region. The function (5–14) can therefore be considered as an extension of the function (5–15) from the unit circle to the entire complex plane with the point $z = 1$ excluded. Indeed, whenever an analytic function is defined by some expression (such as a power series) in some domain D which is not the whole complex plane, it is natural to ask if this function can be extended to a larger domain.

First, consider the function $w_1(z)$ defined on the domain D_1 and let D_2 be another domain, part of which coincides with D_1 as shown in figure 5–6. It can be shown that the common region D_c, which is part of both domains, is itself a domain [58] and hence is a subdomain of both the domain D_1 and the domain D_2. Now suppose that there exists a function w_2 which is analytic in the domain D_2 and which is equal to $w_1(z)$ at every point z of the common domain D_c. Of course, the function $w_2(z)$ may not exist. However, if it does exist, it is called the *direct analytic continuation of the function* w₁ *to the domain* D₂.

There can be at most one direct analytic continuation of a function to any given domain. For if ζ_2 is another direct analytic continuation of w_1 to D_2, then w_2 and ζ_2 are both analytic in D_2 and are equal to one another in the subdomain D_c. Hence, the fundamental theorem shows that w_2 and ζ_2 are equal at every point of D_2. And this of course means that ζ_2 and w_2 are the same function.

---

[58] See, e.g., ref. 8, ch. 1. The domains D_1 and D_2 are said to intersect; and the common domain D_c, called the intersection of D_1 and D_2, is denoted by $D_1 \cap D_2$.

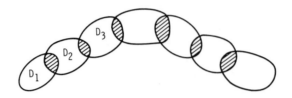

FIGURE 5–7.—Chain of domains for analytic continuation.

Let $\tilde{D}$ be the domain which consists of all points which belong either to the domain D_1 or the domain D_2, or both.[59] Then since the analytic functions $w_1(z)$ and $w_2(z)$ are equal at all points where they are both defined, we can define a new analytic function $\tilde{w}(z)$ on the domain $\tilde{D}$ by the relation

$$\tilde{w}(z) = \begin{cases} w_1(z) & \text{for } z \text{ in } D_1 \\[2ex] w_2(z) & \text{for } z \text{ in } D_2 \end{cases}$$

It is clear that the analytic function $\tilde{w}(z)$ is an extension of the analytic function $w_1(z)$ from the domain D_1 to the larger domain $\tilde{D}$.

The process described above does not have to terminate with the function $w_2(z)$. It may, for example, be possible to find a direct analytic continuation $w_3(z)$ of the function $w_2(z)$ to a domain D_3, and so on. Proceeding in this manner we obtain a chain of domains $D_2, D_3, \ldots$ such as that shown in figure 5–7 and a collection of analytic functions $w_2(z), w_3(z), \ldots$ defined on these domains. Each of these functions is said to be an *analytic continuation* of the function $w_1(z)$, and the procedure itself is called *analytic continuation*. We say that $w_1(z)$ is *analytically continued along a simple* [60] *curve* Γ which extends from D_1 to some point P if Γ is completely covered by a chain of domains $D_2, D_3, \ldots$ (as shown in fig. 5–8) along which $w_1(z)$ can be analytically continued in the manner described above.

---

[59] It is shown in various books on analysis that this extended region is indeed a domain. It is called the union of the domains D_1 and D_2 and is denoted by $D_1 \cup D_2$, e.g., see ref. 8, ch. 1.

[60] Roughly, this means that Γ is smooth and does not cross itself nor have any other pathological behavior.

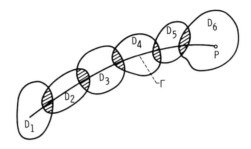

FIGURE 5–8. — Analytic continuation along a curve.

The collection of functions generated by the process of analytic continuation can be used, as in the case of direct analytic continuations, to define a new analytic function on the larger domain which includes all the domains on which the analytic continuations are defined. If, in addition, the function $w_1(z)$ is analytically continued along a curve Γ, the values of this extended function on Γ itself will be the same no matter which specific collection of domains $D_2, D_3, \ldots$ is used to construct it.

5.7.2 Specific Method

In order to make these ideas more concrete we shall consider a specific process which can be used, at least in theory, to obtain an analytic continuation of any given function. Thus, suppose that the analytic function $w_1(z)$ is defined by some expression in the domain D_1. For example, it may be defined by a Taylor series

$$w_1(z) = \sum_{n=0}^{\infty} a_n(z - z_1)^n \qquad (5\text{–}16)$$

in which case the domain D_1 will be the interior of a circle of radius R_1 centered at the point z_1. We suppose that R_1 is finite. Choose a point z_2 in the domain D_1. Since an analytic function is infinitely differentiable, we can calculate the sequence of derivatives $w_1^{(n)}(z_2)$ for $n = 0, 1, 2, \ldots$ from the given expression for $w_1(z)$ in the domain D_1. For example, when the function $w_1(z)$ is given by the Taylor series (5–16), $w_1^{(n)}(z_2)$ can be obtained by differentiating equation (5–15), term by term, n times and evaluating the result at z_2. Then as indicated in section 5.5 the Taylor series

135

$$w_2(z) = \sum_{n=0}^{\infty} b_n(z-z_2)^n$$

with

$$b_n \equiv \frac{w_1^{(n)}(z_2)}{n!} \qquad \text{for } n = 0, 1, 2, \ldots$$

will converge to an analytic function in a circle of nonzero radius centered at z_2. And this function will be equal to $w_1(z)$ at every point which is inside both this circle and the domain D_1. Of course, the circle of convergence of $w_2(z)$ may not extend beyond the domain D_1 (as shown in fig. 5–9(a)). If this occurs,

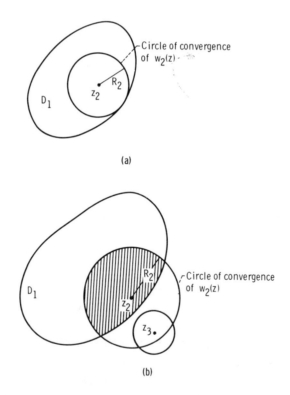

(a)

(b)

(a) Circle of convergence of $w_2(z)$ does not extend beyond D_1.
(b) Circle of convergence of $w_2(z)$ extends beyond D_1.

FIGURE 5–9.—Analytic continuation by power series.

we can choose a new point z_2 and repeat the process. However, if the circle of convergence of $w_2(z)$ does extend beyond D_1 (as shown in fig. 5–9(b)), then $w_2(z)$ will be an analytic continuation of $w_1(z)$. The process can now be repeated by choosing a point z_3 which lies within the circle of convergence of $w_2(z)$, proceeding to obtain a new Taylor series about this point, and so on.

It can be shown that any analytic continuation of a given function, no matter how it has been obtained, can also be found by using the method of power series just described. It is easy to verify from this that the analytic continuation of the derivative of an analytic function has the same value at any given point as the derivative of the analytic continuation at that point, provided they are both carried out along the same curve. This means that the order of differentiation and analytic continuation can be interchanged.

5.7.3 Singular Points

Let $w(z)$ be analytic at all points of a domain D except for a certain number of isolated singular points. Suppose, in addition, that $w(z)$ is analytic on a subdomain D_1 of D and that $w_1(z)$ is the restriction[61] of $w(z)$ to D_1. Then $w_1(z)$ is analytic on D_1 and can be analytically continued to any other subdomain of D which does not contain singular points of $w(z)$. As long as the analytic continuation of $w_1(z)$ is carried out along a path which lies entirely within D, the value of this analytic continuation at any point of D will be equal to the value of $w(z)$ at that point. However, since the circle of convergence of a power-series expansion of an analytic function will pass through its nearest isolated singular point (provided that point is nearer than the boundary of the domain), it can be seen by using the method of power series that the function $w_1(z)$ cannot be analytically continued along any curve which passes through an isolated singular point of $w(z)$. (These ideas are illustrated in fig. 5–10.)

More generally, let $w_1(z)$ be analytic on some domain D_1. If this function cannot be analytically continued along any simple curve which crosses the boundary of D_1 at the point z_0, we say that the point z_0 is a *singular point* of the function $w_1(z)$. And the preceding remarks show that this definition is consistent with the definition of an isolated singular point given in section 5.4.

Let $\sum\limits_{n=0}^{\infty} a_n (z - z_0)^n$ be a power series whose radius of convergence is not

---

[61] That is, $w_1(z)$ is a function which is defined only on D_1 and takes on the same values at each point of D_1 as the function $w(z)$.

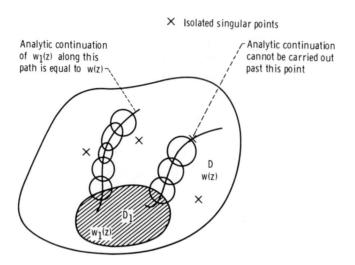

FIGURE 5-10. — Illustration of analytic continuation of a restriction of an analytic function.

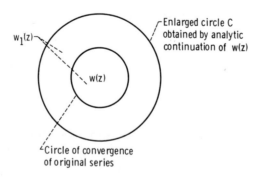

FIGURE 5-11. — Analytic continuation past a circle of convergence.

equal to zero or infinity. We have seen that this series converges to an analytic function $w(z)$ everywhere within its circle of convergence and diverges outside this circle. We shall now show that this circle always passes through a singular point [62] of $w(z)$. In order to obtain a contradiction, suppose that there were no singularities of $w(z)$ on the circle of convergence. Then it would be possible to analytically continue $w(z)$ a finite distance outside this circle everywhere around its circumference (as shown in fig. 5–11). These analytic continuations could then be used to construct an analytic function $w_1(z)$ which is an extension of $w(z)$ to a larger circle C which is also centered at x_0. Now it is shown in section 5.5 that the circle of convergence of the Taylor series expansion of $w_1(z)$ about z_0 cannot be smaller than C. But it is also shown in that section that this extended function $w_1(z)$ must have the same Taylor series expansion about z_0 as $w(z)$. However, this is impossible since (by hypothesis) this latter series diverges outside of the smaller circle. Hence, we must conclude that there is a singular point of $w(z)$ on its circle of convergence.

Starting with a given analytic function $w_1(z)$ defined on a domain D_1, we can carry out the process of analytic continuation until all possible analytic continuations of the function $w_1(z)$ have been found. The collection of analytic functions generated in this manner can again be used to define a new function on the domain which consists of all the domains on which these various functions are defined. The function obtained in this manner is called a *complete analytic function*. This function cannot be further extended. A point z_0 is said to be a singular point of the complete analytic function if it is a singular point of any analytic continuation of $w_1(z)$. And sometimes, when no confusion is likely to arise, we shall say that z_0 is a singular point of the original function $w_1(z)$ itself.

5.7.4 Multiple-Valued Functions

There is a certain difficulty associated with the definition of a complete analytic function given in the preceding section. Thus, suppose that the analytic function $w_1(z)$ defined on the domain D_1 can be analytically continued along the two simple curves Γ_1 and Γ_2 which terminate at the same point p,

---

[62] In section 5.5 we only asserted that the radius of convergence does not exceed the distance between z_0 and the nearest *isolated* singular point.

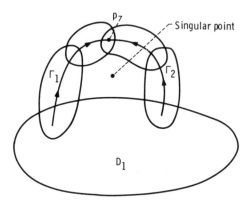

FIGURE 5–12.—Paths for analytic continuation of multiple-valued function.

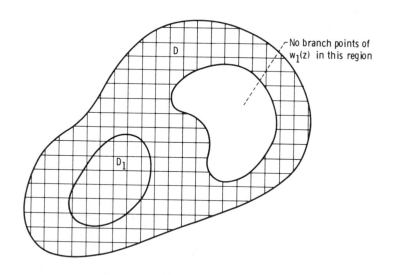

FIGURE 5–13.—Domain for analytic continuation.

as shown in figure 5–12. There is no guarantee that the analytic continuation of $w_1(z)$ along Γ_1 will have the same value at p as its analytic continuation along Γ_2. But if this occurs, the complete analytic function obtained from $w_1(z)$ must have more than one value at the point p. Thus, although up to this point we have assumed that all functions are single valued, we must in general allow a complete analytic function to be multiple valued. It can be shown (ref. 19, p. 218) that the analytic continuation along the path Γ_1 will always have the same value at the point p as the analytic continuation along Γ_2 unless there is a singular point (such as that shown in fig. 5–12) between these two curves.[63] However, the mere existence of a singular point between the two curves does not guarantee that the analytic continuations along the two different curves will have different values at p. This only occurs when the singular point is a *branch point*. A branch point is a singular point of a function which has the property that the function will not return to its starting value upon analytic continuation along any arbitrarily small circle which surrounds this point.

Suppose that the single-valued analytic function $w_1(z)$ is defined on a subdomain D_1 of a domain D (see fig. 5–13). And suppose that D neither contains any singular points of $w_1(z)$ nor is it possible to construct a closed curve within D which surrounds a branch point $w_1(z)$. Then it is impossible for any two analytic continuations of $w_1(z)$ along paths which lie entirely within D to have different values at any given point p of D. Now, in terms of these analytic continuations, we can construct on D (in the manner described above) an analytic function $w(z)$. Then this function will be *single valued*. It is an extension of $w_1(z)$ from the domain D_1 to the larger domain D. And the fundamental theorem shows that $w(z)$ is the only single-valued analytic function with this property. Therefore, when no confusion is likely to arise, we do not distinguish between the two functions $w(z)$ and $w_1(z)$ and we simply say that $w_1(z)$ is defined on the larger domain D.

For example, consider the function $w_1(z)$ defined by

$$w_1(z) = z^{1/2} \tag{5–17}$$

---

[63] This result is known as the *monodromy theorem*.

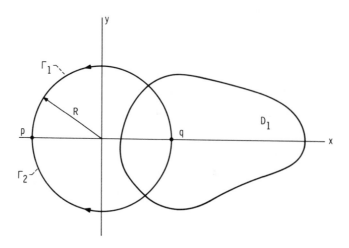

FIGURE 5–14. — Analytic continuation of function $z^{1/2}$.

on a domain D_1 which includes a portion of the real axis but which includes neither the origin nor any portion of the negative real axis, as shown in figure 5–14. It is easy to verify that equation (5–17) represents a single-valued analytic function in the domain D_1. In order to proceed it is convenient to introduce the polar representation $z = re^{i\theta}$ discussed in section 5.1. Then the function (5–17) can be written as[64]

$$w_1(re^{i\theta}) = r^{1/2}e^{i\theta/2} \tag{5–18}$$

When the point z is in D_1, the argument θ will always lie in the range

$$-\pi < \theta < \pi \tag{5–19}$$

(Notice that strict inequality signs are used.) Since the formula (5–18) with an extended range of θ determines an analytic function at each finite point of the complex plane except $r = 0$ and since this function coincides with $w_1(z)$ in D_1, we can use this formula to analytically continue $w_1(z)$ outside of D_1.

---

[64] Recall that we have adopted the convention that the square root of a positive real number is always the positive square root.

Now the origin is a singluar point of $w_1(z)$. Hence, suppose we analytically continue this function to the point p along Γ_1, the semicircle with radius R in the upper half plane shown in figure 5–14. Then the value of this analytic continuation at p is $R^{1/2}e^{i(\pi/2)} = R^{1/2}[\cos(\pi/2) + i \sin(\pi/2)] = iR^{1/2}$. But the value at p of the analytic continuation along the semicircle Γ_2 is $R^{1/2}e^{-i(\pi/2)} = R^{1/2}[\cos(-\pi/2) + i \sin(-\pi/2)] = -iR^{1/2}$. These analytic continuations of the function $w_1(z)$ therefore have different values at the point p.

If, instead of stopping at the point p, we carry out the analytic continuation of $w_1(z)$ first along Γ_1 to the point p and then along Γ_2 from the point p to the point q (in the direction opposite to the arrows), we arrive at the value $R^{1/2}e^{i2\pi/2} = -R^{1/2}$. But since the original value of $w_1(z)$ at the point q is $R^{1/2}e^{i0} = R^{1/2}$, we see that the function does not return to its original value upon analytic continuation around this circle. And since the radius R is arbitrary, this shows that the origin is a branch point of $w_1(z)$. By making the transformation $z = 1/w$ and taking R arbitrarily large, we can also show that the point at infinity is a branch point of this function. And since $z = 0$ and $z = \infty$ are the only singular points (and therefore the only branch points) of $w_1(z)$, this function will always return to its original value when it is analytically continued around any path which does not enclose the origin.[65]

Now every analytic continuation of $w_1(z)$ can be obtained from the formula (5–18) by letting r range between zero and infinity and letting θ take on all values both positive and negative. But, since $e^{2in\pi} = 1$ for $n = 0, \pm 1, \pm 2, \ldots$, we need only consider values of θ in the range $\theta_0 \leqslant \theta < \theta_0 + 4\pi$ (where θ_0 can be chosen as any fixed number) in order to obtain all possible values of the function (5–18). Thus, the complete analytic function $\tilde{w}(z)$ obtained from $w_1(z)$ is the multiple-valued function defined by

$$\tilde{w}(re^{i\theta}) = r^{1/2}e^{i\theta/2} \qquad 0 \leqslant r \leqslant \infty; \; \theta_0 \leqslant \theta \leqslant \theta_0 + 4\pi \qquad (5\text{–}20)$$

It takes on two distinct values at each finite point of the complex plane except at the origin, which is a branch point. And one of these two values is equal to the negative of the other.

---

[65] It will never return to its original value when analytically continued once around any path which does enclose the origin.

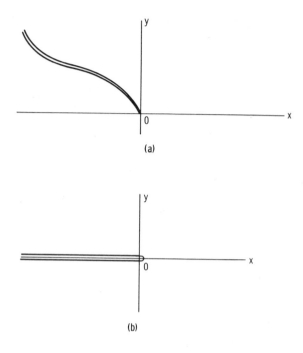

(a) Branch cut in arbitrary location.
(b) Branch cut along negative real axis.

FIGURE 5–15.—Branch cuts for $z^{1/2}$.

It is usually undesirable to deal with multiple-valued functions.[66] We can avoid doing this by drawing a line connecting the branch point of the function $z^{1/2}$ at the origin with its branch point at infinity, as shown in figure 5–15(a), and then restricting the analytic continuations so that they are not carried out along any path which crosses this line. Such a line is called a *branch cut* and it is used to prevent analytic continuations from being carried out along curves which encircle the origin. The actual location of the branch cut is arbitrary but we may, for definiteness, assume that it lies along the negative real axis, as shown in figure 5–15(b). Then starting with the original function (5–17) (which, as can be seen from eqs. (5–18) and (5–19), is positive along the positive real axis) and analytically continuing this function along all allowable paths in the complex plane, we obtain the extended function

---

[66] After all, we would not expect a well-defined physical problem to have a solution which is multiple valued.

$$W_1(re^{i\theta}) = r^{1/2}e^{i\theta/2} \qquad 0 \leq r \leq \infty; \; -\pi \leq \theta \leq \pi$$

which is single valued and analytic at every point of the complex plane not lying on the negative real axis. We can also analytically continue the function which is equal to the negative of the original function (5–17) in D_1 along all allowable paths in the cut plane to obtain the extended function

$$W_2(re^{i\theta}) = r^{1/2}e^{i\theta/2} \qquad 0 < r < \infty; \; \pi < \theta < 3\pi$$

And this function is also single valued and analytic at every point of the complex plane not lying on the negative real axis. The functions W_1 and W_2 are said to be *branches* of the double-valued function (5–20). Taken together these two branches assume all the values of the multiple-valued function and are therefore equivalent to it. Hence, we can deal with a multiple-valued function by replacing it with its single-valued branches.

The complete function in this example is double valued and therefore has two branches. However, we also encounter multiple-valued functions which take on infinitely many values at each point and therefore have infinitely many branches. For example, the function $w(z) = \ln z$ is defined in polar notation to be

$$w(re^{i\theta}) = \ln re^{i\theta} = \ln r + i\theta$$

In order to obtain the complete analytic function we must (as in the preceding example) let r take on all values in the range $0 < r < \infty$ and θ all real values. But since $z = re^{i\theta} = re^{i(\theta + 2n\pi)}$ if, and only if, $n = 0, \pm 1, \pm 2, \ldots$, this complete function must have infinitely many values at each point. And these values differ from one another by multiples of $2\pi i$. This function also has a branch point at the origin and a branch point at infinity. And if the branch cut is again taken along the negative real axis, the infinitely many branches $w_n(z)$ *for* $n = 0, \pm 1, \pm 2, \ldots$ of $\ln z$ become

$$w_n(re^{i\theta}) = \ln r + i\theta \qquad 0 < r < \infty; \; (2n-1)\pi < \theta < (2n+1)\pi$$
$$\text{for } n = 0, \pm 1, \pm 2, \ldots$$

The branch corresponding to the range $-\pi < \theta < \pi$ is called the *principal branch* of the logarithm.

The function z^α (where α is some complex constant) can be defined in terms of the logarithm by the formula $z^\alpha = e^{\alpha \ln z}$. Hence, the branch points of this function can be located only at the origin and at infinity. By using polar notation we can express this function in the form $z^\alpha = e^{\alpha \ln r} e^{\alpha i\theta} = r^\alpha e^{i\alpha\theta}$. And this shows that the complete function is multiple valued unless α is an integer. In fact, it takes on infinitely many values at each point unless α is a rational number. The various branches of this function can be formed in the same way as for the logarithm.

In a similar manner, it can be seen that for any finite point z_0 the complete analytic function associated with $(z - z_0)^\alpha$ is multiple valued whenever α is not an integer. Its branch points are $z = z_0$ and $z = \infty$ and its branch cut can be taken along any line joining these two points. However, once a branch cut has been chosen, the various branches of this function will then be analytic everywhere in the cut plane.

5.8 PERMANENCE OF FUNCTIONAL RELATIONS

Let $F(z_1, \ldots, z_n)$ be a complex function of the n complex variables $z_1 = x_1 + iy_1, \ldots, z_n = x_n + iy_n$. If this function can be expanded in a power series

$$F(z_1, \ldots, z_n) = \sum_{i_1, \ldots, i_n=1}^{\infty} a_{i_1}, \ldots, a_{i_n} z_1^{i_1} z_2^{i_2} \ldots z_n^{i_n}$$

with complex coefficients $a_{i_1}, \ldots, a_{i_n}$ and if this series converges in some neighborhood of each point in some domain of the $2n$-dimensional space whose coordinates are $x_1, \ldots, x_n, y_1, \ldots, y_n$, then we say that F is an analytic function of the n complex variables $z_1, \ldots, z_n$ in D. This is clearly an extension of the definition of an analytic function of a single complex variable given in section 5.2.

Now let $F(w_1, \ldots, w_n, z)$ be an analytic function of the $n+1$ complex variables $w_1, \ldots, w_n, z$ for all values of the variables $w_1, \ldots, w_n$ and for all values of z in some domain D_0. If $w_1(z), \ldots, w_n(z)$ are analytic functions of the complex variable z (in the usual sense) in some common subdomain D of D_0, then it can be shown (ref. 4) that the function $g(z)$ defined by

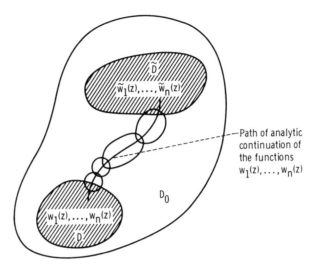

FIGURE 5–16.—Illustration of permanence of functional identities.

$$g(z) \equiv F(w_1(z), \ \ldots, w_n(z), z)$$

is also analytic in D.

Now suppose that $\tilde{w}_1(z), \ \ldots, \ \tilde{w}_n(z)$ are analytic continuations of $w_1(z), \ \ldots, \ w_n(z)$, respectively, from D to some other common subdomain $\tilde{D}$ of D_0 and these analytic continuations can all be obtained by analytically continuing the functions $w_1(z), \ \ldots, \ w_n(z)$ along a single curve Γ which lies entirely within D_0. (This is illustrated in fig. 5–16.) In addition, suppose that $w_1(z), \ \ldots, w_2(z)$ satisfy the equation

$$F(w_1, \ \ldots, w_n, z) = 0 \qquad (5\text{–}21)$$

at all points z in D. Then by analytically continuing these functions along Γ to $\tilde{D}$ and using the fundamental theorem of analytic continuation given in the beginning of section 5.7, it can be shown (ref. 19, p. 210) that $\tilde{w}_1(z), \ \ldots, \tilde{w}_n(z)$ also satisfy equation (5–21) at every point of $\tilde{D}$. This is known as the *principle of permanence of functional relations*. Roughly speaking, it means that the analytic continuations of the solutions of equation (5–21) are also solutions of this equation.

147

5.9 DIFFERENTIAL EQUATIONS IN COMPLEX PLANE

5.9.1 Definition

We have already indicated the utility of extending the definition of a differential equation to include the case where the variables are complex. To this end let $F(w_1, \ldots, w_n, z)$ be an analytic function of the $n+1$ complex variables $w_1, \ldots, w_n, z$ in some domain D. Then the nth-order normal differential equation in the complex domain is an equation of the form

$$\frac{d^n w}{dz^n} = F\left(w, \frac{dw}{dz}, \ldots, \frac{d^{n-1} w}{dz^{n-1}}, z\right) \qquad (5\text{--}22)$$

Notice that in writing this equation we imply that its solutions, if they exist, must possess complex derivatives at all points where they are defined. It is also reasonable to require that the solutions satisfy the equation at least on some domain in the z-plane. Hence, the solutions to equation (5–22) must be analytic functions. This is a much stronger restriction than is imposed in the case of real variables, where we require only that the solutions be sufficiently differentiable.

5.9.2 Fundamental Theorem

The following fundamental theorem (which is analogous to that given in chapter 1 for the real-variable case) can be shown to hold (ref. 4, p. 119). *For each point* [67] $\zeta_1, \ldots, \zeta_n, z_0$ *of the domain D where the function F is analytic, there exists a unique (i.e., single-valued) function w(z) which satisfies the initial conditions*

$$w(z_0) = \zeta_1, \; w'(z_0) = \zeta_2, \; \ldots, \; w^{(n-1)}(z_0) = \zeta_n$$

is analytic, and satisfies equation (5–22) in some neighborhood of the point z_0.

We shall be principally interested in the (effectively normal) linear equation

$$\frac{d^n w}{dz^n} + a_1(z) \frac{d^{n-1} w}{dz^{n-1}} + \ldots + a_{n-1}(z) \frac{dw}{dz} + a_n(z) w + b(z) = 0 \qquad (5\text{--}23)$$

where the coefficients $a_1(z), \ldots, a_n(z), b(z)$ are all analytic on some common domain D_0. This equation is effectively of the form (5–22) with the function F

---

[67] $\zeta_1, \ldots, \zeta_n, z_0$ are $n+1$ complex numbers.

analytic on the domain D which consists of all the values of the variables $w_1, \ldots, w_2$ and all the values of the variable z which lie in the domain D_0. Hence, the fundamental theorem now becomes: *For each set of complex numbers $\zeta_1, \ldots, \zeta_n$ and each point z_0 of D_0 there exists a unique function $w(z)$ which satisfies the initial conditions*

$$w(z_0) = \zeta_1, \; w'(z_0) = \zeta_2, \; \ldots, \; w^{(n-1)}(z_0) = \zeta_n$$

is analytic, and satisfies equation (5–23) in some neighborhood of the point z_0.

There is also, in this case, an additional result, due to Fuchs (ref. 24, p. 4) which asserts that *this solution, $w(z)$, has a Taylor series expansion about z_0 whose radius of convergence is at least equal to the shortest distance between z_0 and the boundary of D_0.*

Now suppose that $w(z)$ is a solution to equation (5–23) on some subdomain D_1 of D_0. Then the (single valued) function $w(z)$ and all its derivatives are analytic on D_1. Let $\tilde{w}(z)$ be an analytic continuation of $w(z)$ along some curve in D_0 to some other subdomain $\tilde{D}$ of D_0, as shown in figure 5–17. Then since, as indicated in section 5.7.2, the analytic continuation of a derivative of an analytic function is equal to the derivative of the analytic continuation, we can apply the principle of permanence of functional relations to equation (5–23)

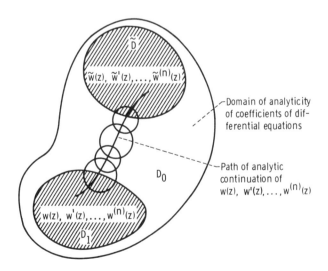

FIGURE 5–17.—Illustration of analytic continuation of solution to differential equation.

to show that the analytic continuation $w(z)$ is itself a solution to equation (5–23). *Thus, the analytic continuation along any curve lying entirely within D_0 of a solution to equation (5–23) is also a solution of this equation.*

For example, the differential equation

$$w'' + \frac{1}{6z}w' + \frac{1}{6z^2}w = 0$$

has the solution

$$w(z) = z^{1/2} + z^{1/3}$$

But upon analytic continuation of this solution around any closed path encircling the origin, we obtain the function

$$w_1(z) = (e^{2\pi i}z)^{1/2} + (e^{2\pi i}z)^{1/3} = -z^{1/2} + e^{2\pi i/3}z^{1/3}$$

And it is easy to verify by direct substitution that $w_1(z)$ is also a solution of the equation. (In fact, $w(z)$ and $w_1(z)$ are linearly independent.)

We shall now show that *any solution to equation (5–23) which is defined on a subdomain of D_0 can indeed be analytically continued along any curve in D_0.* This means that *no solution to equation (5–23) can have a singular point in D_0.* The assertion can be proved by assuming that there exists a solution $w(z)$ in a subdomain D_1 of D_0 which cannot be analytically continued along some curve Γ in D_0 and then showing that this leads to a contradiction. Thus, if $w(z)$ cannot be continued along Γ, there must be a point z_1 on Γ as shown in figure 5–18 such that $w(z)$ can be continued up to, but not past, this point. We can therefore choose a point z_2 on the portion of Γ joining D_1 to z_1 which is closer to z_1 than any part of the boundary of D_0. Upon analytically continuing the solution $w(z)$ along Γ to z_2, we obtain a solution $w_2(z)$ of equation (5–23) in a neighborhood of z_2. And Fuchs' result shows that $w_2(z)$ can be expanded in a Taylor series about z_2 that converges in a circle which includes the point z_1. But this constitutes an analytic continuation of $w(z)$ along Γ past the point z_1, which was assumed to be impossible. And this proves the assertion.

It follows from these results that any solution [68] to equation (5–23) in

---

[68] The existence of such a solution is asserted by the fundamental theorem but only in some neighborhood of this point. And this neighborhood could be very small.

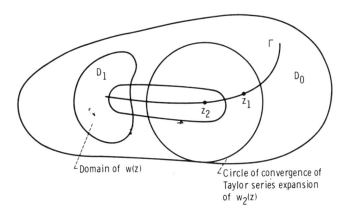

FIGURE 5–18.—Continuation of $w(z)$ along Γ.

a neighborhood of a point where the coefficients are analytic can actually be extended to obtain a solution to this equation on the entire domain D_0 on which the coefficients are analytic. However, in order to do this we must, in general, allow these solutions to be multiple valued.

5.9.3 Linearly Independent Solutions

By using the fundamental theorem given in this section, the various results given in section 1.6 for linear equations with real variables can be extended to the complex-variable case. Thus, the homogeneous equation associated with equation (5–23)

$$\frac{d^n w}{dz^n} + a_1(z)\frac{d^{n-1}w}{dz^{n-1}} + \ldots + a_{n-1}(z)w = 0 \qquad (5-24)$$

possesses n linearly independent (single valued) solutions $w_1(z), \ldots, w_n(z)$, called a *fundamental set* of solutions, in a neighborhood of each point of D_0. Now let $w_p(z)$ be a particular (single valued) solution of equation (5–23) in the neighborhood of some point of D_0. Then every solution of this equation about this point can be obtained by making a suitable choice of the arbitrary constants $c_1, \ldots, c_n$ in the general solution $w(z) = c_1 w_1(z) + \ldots + c_n w_n(z) + w_p(z)$ of this equation.

The definition and discussion of linear independence given in section 1.6 applies with only trivial modification to the case where the functions are complex and analytic. Thus, in particular, a necessary and sufficient condition (ref. 24, p. 13) that n (single valued) solutions $w_1(z), \ldots, w_n(z)$ of equation (5–24) be linearly independent on a domain D_0 on which the coefficients are analytic is that the Wronskian

$$W(w_1, \ldots, w_n) \equiv \begin{vmatrix} w_1 & w_2 & \cdots & w_n \\ w_1' & w_2' & \cdots & w_n' \\ \cdot & & & \\ \cdot & & & \\ \cdot & & & \\ w_1^{(n-1)} & w_2^{(n-1)} & \cdots & w_n^{(n-1)} \end{vmatrix} \quad (5\text{–}25)$$

not be equal to zero at any point z_1 of D_0. Since $w_1, \ldots, w_n$ are analytic functions of z, the Wronskian itself is an analytic function of z which we shall denote by $\mathscr{W}(z)$. Thus,

$$\mathscr{W}(z) \equiv W(w_1(z), \ldots, w_n(z))$$

It is also easy to see that the Wronskian of the analytic continuations of $w_1(z), \ldots, w_n(z)$ along a curve Γ in D_0 is equal to the analytic continuation of the function $\mathscr{W}(z)$ along Γ.

It can be shown by using the rules for differentiating determinants and by substituting in the differential equation (5–24) (ref. 24, p. 12) that $\mathscr{W}$ satisfies the first-order differential equation

$$\frac{d\mathscr{W}}{dz} = -p_1(z)\mathscr{W} \quad (5\text{–}26)$$

And upon separating the variables and integrating along any path Γ in D_0, we obtain

$$\mathscr{W}(z) = ce^{-\int p_1(z)dz} \quad (5\text{–}27)$$

where c is a complex constant of integration.

Since $p_1(z)$ is analytic in D_0, its integral $\int p_1(z)dz$ must also be analytic in D_0. Hence, in particular, this integral cannot become infinite at any point of D_0; and therefore the exponential factor in equation (5–27) can never vanish in D_0. Thus, $\mathscr{W}(z)$ can only be equal to zero at a point z_1 of D_0 if $c=0$. And this shows that, if $\mathscr{W}(z)$ vanishes at any point of D_0, it must vanish at every point of D_0.

In the special case where $n=2$, equation (5–27) can be used to obtain an explicit formula which determines a second linearly independent solution $w_2(z)$ to equation (5–24) when one solution $w_1(z)$ to this equation is known. Thus, when $n=2$, we find upon expanding the determinant and rearranging that

$$w_1^2 \frac{d}{dz}\left(\frac{w_2}{w_1}\right) = \mathscr{W}(z) = ce^{-\int p_1(z)dz}$$

And integrating this along any curve in D_0 yields the formula

$$w_2(z) = cw_1(z)\int \frac{e^{-\int p_1(z)dz}}{[w_1(z)]^2}\,dz \tag{5–28}$$

which agrees with the formula obtained by the method of variation of parameters in section 4.1. It follows from the way in which it was constructed that any solution $w_2(z)$ calculated from this formula will be linearly independent of $w_1(z)$.

More generally, it can be shown (ref. 24, p. 16, example 10) that, if $n-1$ linearly independent solutions, say $w_1(z), \ldots, w_{n-1}(z)$, to equation (5–24) are known, another linearly independent solution to this equation is given by

$$w_n = c\sum_{i=1}^{n-1} w_i(z)\int \frac{e^{-\int p_1(z)dz}}{[\mathscr{W}_1(z)]^2}\,M_i(z)dz$$

where $\mathscr{W}_1(z)$ is the Wronskian

153

$$\mathscr{W}_1(z) = W(w_1, \ldots, w_{n-1}) \equiv \begin{vmatrix} w_1 & w_2 & \cdots & w_{n-1} \\ w_1' & w_2' & \cdots & w_{n-1}' \\ \vdots & & & \\ w_1^{(n-2)} & w_2^{(n-2)} & \cdots & w_{n-1}^{(n-2)} \end{vmatrix}$$

and, for $i = 1, 2, \ldots, n-1$, $M_i(z)$ is the cofactor of w_i^{n-2} in this Wronskian. That is,

$$M_i(z) = \frac{\partial W}{\partial w_i^{n-1}}$$

5.10   NONELEMENTARY TRANSCENDENTAL FUNCTIONS

We shall have occasion to use two particular nonelementary functions of a complex variable called the *gamma* function and the *beta* function. First, we define the analytic function $\Gamma(z)$ for $\mathscr{R}e\ z > 0$ to be the *Eulerian integral of the second kind.*

$$\Gamma(z) = \int_0^\infty e^{-t} t^{z-1} dt \tag{5–29}$$

This integral converges in the right half plane $\mathscr{R}e\ z > 0$ and diverges for $\mathscr{R}e\ z \leqslant 0$. It can be shown (ref. 25) that it represents an analytic function in its domain of convergence. Although this analytic function is only defined by equation (5–29) in the right half plane, it can be analytically continued into the left half plane $\mathscr{R}e\ z \leqslant 0$ by using the formula

$$\Gamma(z)\Gamma(1-z) = \frac{\pi}{\sin \pi z} \tag{5–30}$$

to compute the values of $\Gamma(z)$ for $\mathscr{R}e\ z \leqslant 0$ from its values at points in the right half plane. It follows from this equation that $\Gamma(z)$ has simple poles at $z = 0, -1, -2, \ldots$.

Integrating equation (5–29) by parts (with z replaced by $z+1$) shows that for $\mathscr{R}e\ z > 0$

$$\Gamma(z+1) = z \int_0^\infty e^{-t} t^{z-1} dt$$

Hence,

$$\Gamma(z+1) = z\Gamma(z) \qquad (5\text{–}31)$$

By successively applying equation (5–31) we find that for any positive integer n

$$\Gamma(z+n) = (z+n-1)\Gamma(z+n-1)$$

$$= (z+n-1)(z+n-2)\Gamma(z+n-2)$$

$$\vdots$$

$$= (z+n-1)(z+n-2) \ \cdots \ (z+1)z\Gamma(z) \qquad (5\text{–}32)$$

It is convenient to introduce a special notation for the factor multiplying $\Gamma(z)$ in the last member of this equation. Hence, we define the *generalized factorial function* $(z)_n$ by

$$(z)_0 = 1$$

and

$$(z)_n \equiv \prod_{m=0}^{n-1} (z+m)$$

$$= z(z+1)(z+2) \ \cdots \ (z+n-1) \qquad \text{for } n = 1, 2, 3, \ldots$$

$$(5\text{–}33)$$

Thus, the symbol $(z)_n$ denotes the product of n factors, each factor being one larger than the preceding one. For example,

$$(7)_3 = 7 \times 8 \times 9$$

$$\left(\frac{1}{2}\right)_4 = \frac{1}{2} \times \frac{3}{2} \times \frac{5}{2} \times \frac{7}{2}$$

Notice that when $z = 1$ in equation (5–33), we obtain the ordinary factorial function since

$$(1)_n = 1 \cdot 2 \cdot 3 \ldots n = n! \tag{5–34}$$

By using this notation, equation (5–32) can be rewritten as $\Gamma(z+n) = (z)_n \Gamma(z)$. And, therefore, the generalized factorial function can be expressed in terms of the gamma function by

$$(z)_n = \frac{\Gamma(z+n)}{\Gamma(z)} \qquad \text{for } n = 1, 2, 3, \ldots \tag{5–35}$$

Since integrating equation (5–29) with $z = 1$ shows that $\Gamma(1) = 1$, we find from equations (5–34) and (5–35) that

$$\Gamma(n+1) = n! \qquad \text{for } n = 1, 2, 3, \ldots \tag{5–36}$$

This equation is also valid for $n = 0$ provided we use the usual definition $0! = 1$.

The *beta* function $B(z, \zeta)$ is a function of two complex variables and is defined by

$$B(z, \zeta) \equiv \int_0^1 t^{z-1}(1-t)^{\zeta-1}dt \tag{5–37}$$

for $\mathscr{Re}\, z > 0$ and $\mathscr{Re}\, \zeta > 0$. However, it is possible to express this function in terms of the gamma function since

$$\Gamma(z)\Gamma(\zeta) = \left(\int_0^\infty e^{-t}t^{z-1}dt \right)\left(\int_0^\infty e^{-\tau}\tau^{\zeta-1}d\tau \right) = \int_0^\infty e^{-t}t^{z-1}\left(\int_0^\infty e^{-\tau}\tau^{\zeta-1}d\tau \right) dt$$

And upon setting $x = \tau/t$, we can eliminate τ from this equation to obtain

$$\Gamma(z)\Gamma(\zeta) = \int_0^\infty e^{-t}t^{z-1}t^{\zeta}\left(\int_0^\infty e^{-tx}x^{\zeta-1}dx \right) dt$$

Hence, after changing the order of integration we find that

$$\Gamma(z)\Gamma(\zeta) = \int_0^\infty x^{\zeta-1}\left(\int_0^\infty e^{-t(x+1)}t^{z+\zeta-1}dt \right) dx$$

Then by defining τ_1 by $\tau_1 = t(x+1)$ and eliminating t, we get

$$\Gamma(z)\Gamma(\zeta) = \int_0^\infty \frac{x^{\zeta-1}}{(1+x)^{\zeta+z}}\, dx \int_0^\infty e^{-\tau_1}\tau_1^{z+\zeta-1}d\tau_1 \;=\; \Gamma(z+\zeta)\int_0^\infty \frac{x^{\zeta-1}}{(1+x)^{\zeta+z}}\, dx$$

And after setting $x = t/(1-t)$ this becomes

$$\Gamma(z)\,\Gamma(\zeta) = \Gamma(z+\zeta)\int_0^1 t^{\zeta-1}(1-t)^{z-1}dt$$

But comparing this with equation (5–32) shows that

$$B(\zeta, z) = \frac{\Gamma(x)\Gamma(\zeta)}{\Gamma(z+\zeta)}$$

CHAPTER 6

Solution of Linear Second-Order Differential Equations in the Complex Plane

In section 5.9 we extended the definition of a differential equation to include the case where the variables are complex. In this chapter the ideas presented in that section are used to find the behavior of and, in certain cases, to construct solutions to the second-order differential equation

$$\frac{d^2w}{dz^2} + p(z)\,\frac{dw}{dz} + q(z)w = 0 \qquad (6\text{--}1)$$

where z is a complex variable and the coefficients $p(z)$ and $q(z)$ are analytic functions in some domain D of the z-plane. We have seen that every solution of this equation is an analytic function of z.

A point at which the coefficients of equation (6–1) are both analytic is called an *ordinary point* of this equation. Thus, every point of D is an ordinary point. A point z_0 which is a singular point of either $p(z)$ or $q(z)$ (or both) is said to be a *singular point*[69] of equation (6–1). And if the only singularities of $p(z)$ and $q(z)$ which occur at z_0 are isolated singular points, z_0 is also said to be an *isolated singular point* of equation (6–1).

6.1 GENERAL BEHAVIOR OF SOLUTIONS AT ORDINARY POINTS

Let z_1 be any point of D. It was indicated in section 5.9 that equation (6–1) will possess two linearly independent (single valued) solutions, $w_1(z)$ and $w_2(z)$, in some neighborhood of z_1. These solutions are said to be a *fundamental set* of solutions. They possess Taylor series expansions about z_1, say

---

[69] Notice that this definition is consistent with the one given in section 1.5.

159

$$w_1(z) = \sum_{n=0}^{\infty} a_n^{(1)}(z - z_1)^n$$

$$w_2(z) = \sum_{n=0}^{\infty} a_n^{(2)}(z - z_1)^n$$

(6-2)

whose radii of convergence are at least equal to the shortest distance between z_1 and the boundary of D. Since the coefficients of equation (6-1) can always be analytically continued across those points on the boundary of D which are not singular points of the equation, it is always possible to extend the domain D (in which the coefficients are analytic) in such a way that every finite point on its boundary[70] is a singular point of equation (6-1). Hence, we can assert that *the radii of convergence of the Taylor series expansions (6-2) of the solutions $w_1(z)$ and $w_2(z)$ are at least equal to the distance R_0 between z_1 and the nearest singular point of equation (6-1)*. It was also shown in section 5.9 that *the solutions $w_1(z)$ and $w_2(z)$ can be analytically continued along any simple curve in D and that these analytic continuations are themselves solutions of equation (6-1)*. Thus, in particular, we can assert that the series (6-2) not only converge within a circle of radius R_0 but that they also satisfy the differential equation everywhere within this circle.

6.2 GENERAL BEHAVIOR OF SOLUTIONS NEAR ISOLATED SINGULAR POINT

Suppose that z_0 is either an isolated singular point or an ordinary point of equation (6-1). Then the coefficients $p(z)$ and $q(z)$ of this equation will be analytic within the punctured circular domain $0 < |z - z_0| < R$ (shown in fig. 6-1) whose radius R is equal to the distance between z_0 and the singular point [71] of equation (6-1) closest to z_0. We shall call this domain [72] Δ. Now let $w_1(z)$ and $w_2(z)$ be a fundamental set of solutions to equation (6-1) in a neighborhood D_0 of a point z of Δ. We have seen in section 5.9 that these solutions can be extended (by using their analytic continuations) so that they satisfy equation (6-1) at every point of Δ. However, since z_0 may be a singular point and since it can be encircled by a curve such as the curve Γ shown in figure 6-1, these extended solutions can be multiple valued.

---

[70] Provided D has a boundary.

[71] Other than z_0 itself.

[72] Notice that z_0 is *not* a point of Δ.

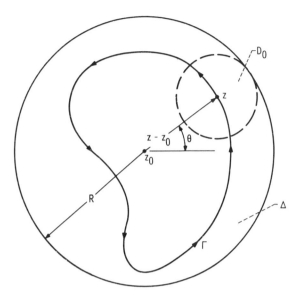

FIGURE 6–1.—Contour for analytic continuation of w_1 and w_2.

In order to deduce the behavior of these extended solutions, we shall now construct a fundamental set of solutions on Δ whose structure is particularly transparent. To this end let the fundamental set $w_1(z)$ and $w_2(z)$ be analytically continued from D_0 counterclockwise around Γ. This process will, in general, yield two new functions of z, say $W_1(z)$ and $W_2(z)$ which are defined on D_0 by

$$\left.\begin{array}{l} W_1(z) = w_1\big(e^{2\pi i}(z-z_0)+z_0\big) \\[2mm] W_2(z) = w_2\big(e^{2\pi i}(z-z_0)+z_0\big) \end{array}\right\} \tag{6-3}$$

But we know that $W_1(z)$ and $W_2(z)$ must satisfy the differential equation (6–1) and that every solution to this equation in D_0 can be expressed as a linear combination of the fundamental set of solutions $w_1(z)$ and $w_2(z)$. Thus, there exist complex constants $a_{11}, a_{12}, a_{21}, a_{22}$ such that

$$\left.\begin{array}{l} W_1(z) = a_{11}w_1(z)+a_{12}w_2(z) \\[2mm] W_2(z) = a_{21}w_1(z)+a_{22}w_2(z) \end{array}\right\} \tag{6-4}$$

161

For example, consider the differential equation [73]

$$w'' + \frac{1}{6z} w' + \frac{1}{6z^2} w = 0$$

and let $z_0 = 0$. A fundamental set of solutions to this equation is $w_1 = z^{1/2}$ and $w_2 = z^{1/3}$. Upon applying equations (6–3) to these solutions, we find that

$$W_1(z) = w_1(e^{2\pi i}z) = (e^{2\pi i}z)^{1/2} = (-1)w_1(z)$$

and

$$W_2(z) = e^{2\pi i/3}z^{1/3} = e^{2\pi i/3}w_2(z)$$

And, therefore, the constants which appear in equations (6–4) become, in this case, $a_{11} = -1$, $a_{12} = 0$, $a_{21} = 0$, $a_{22} = e^{2\pi i/3}$.

We shall now show (in the general case) how the constants in equations (6–4) can be used to construct the desired fundamental set of solutions to equation (6–1) on the domain Δ. In order to do this, we first recall from the theory of linear equations that the algebraic equations

$$\left.\begin{array}{c} (a_{11} - \lambda)c_1 + a_{21}c_2 = 0 \\ a_{12}c_1 + (a_{22} - \lambda)c_2 = 0 \end{array}\right\} \tag{6–5}$$

have a nontrivial solution[74] in c_1, c_2, provided λ is a root of the *characteristic equation*

$$\begin{vmatrix} a_{11} - \lambda & a_{21} \\ a_{12} & a_{22} - \lambda \end{vmatrix} = \lambda^2 - (a_{11} + a_{22})\lambda + (a_{11}a_{22} - a_{12}a_{21}) = 0 \tag{6–6}$$

This equation will either have two distinct roots or two equal roots. We shall treat these two cases separately.

---

[73] This equation was discussed in the example given in section 5.9.2.
[74] That is, a solution other than $c_1 = c_2 = 0$.

6.2.1 Case I: Roots of Characteristic Equation Distinct

First, suppose that the two roots of equation (6–6), λ_1 and λ_2, are distinct. Let $c_1^{(1)}$, $c_2^{(1)}$ be a set of nontrivial solutions to equations (6–5) corresponding to the root λ_1 and $c_1^{(2)}$, $c_2^{(2)}$ be a set of nontrivial solutions corresponding to the root λ_2. Thus,

$$\left. \begin{aligned} (a_{11} - \lambda_1)c_1^{(1)} + a_{21}c_2^{(1)} &= 0 \\[2mm] a_{12}c_1^{(1)} + (a_{22} - \lambda_1)c_2^{(1)} &= 0 \end{aligned} \right\} \tag{6--7a}$$

and

$$\left. \begin{aligned} (a_{11} - \lambda_2)c_1^{(2)} + a_{21}c_2^{(2)} &= 0 \\[2mm] a_{12}c_1^{(2)} + (a_{22} - \lambda_2)c_2^{(2)} &= 0 \end{aligned} \right\} \tag{6--7b}$$

Now let $u_1(z)$ and $u_2(z)$ be the solutions to equation (6–1) which are defined in D_0 (and only in D_0) in terms of the fundamental set $w_1(z)$ and $w_2(z)$ by

$$u_1(z) = c_1^{(1)}w_1(z) + c_2^{(1)}w_2(z) \tag{6--8}$$

$$u_2(z) = c_1^{(2)}w_1(z) + c_2^{(2)}w_2(z) \tag{6--9}$$

Then it follows from equations (6–3), (6–4), and (6–8) that $u_1(e^{2\pi i}(z-z_0)+z_0)$, the analytic continuation of $u_1(z)$ counterclockwise around Γ, is given by

$$u_1(e^{2\pi i}(z - z_0) + z_0) = c_1^{(1)}w_1(e^{2\pi i}(z - z_0) + z_0) + c_2^{(1)}w_2(e^{2\pi i}(z - z_0) + z_0)$$

$$= c_1^{(1)}W_1(z) + c_2^{(1)}W_2(z)$$

$$= [c_1^{(1)}a_{11} + c_2^{(1)}a_{21}]w_1(z) + [c_1^{(1)}a_{12} + c_2^{(1)}a_{22}]w_2(z)$$

Hence, equations (6–7a) and (6–8) imply

$$u_1(e^{2\pi i}(z - z_0) + z_0) = \lambda_1 c_1^{(1)}w_1(z) + \lambda_1 c_2^{(1)}w_2(z) = \lambda_1 u_1(z) \tag{6--10}$$

163

And the same argument shows that the analytic continuation of the solution $u_2(z)$ around Γ, $u_2(e^{2\pi i}(z - z_0) + z_0)$, is

$$u_2(e^{2\pi i}(z - z_0) + z_0) = \lambda_2 u_2(z) \tag{6-11}$$

The constant λ_1 is nonzero. If this were not true, equation (6–10) would show that $u_1(e^{2\pi i}(z - z_0) + z_0)$ is identically zero in D_0. But then $u_1(e^{2\pi i}(z - z_0) + z_0)$ could be analytically continued backwards along Γ to show that $u_1(z)$ is identically zero. And since the constants $c_1^{(1)}$ and $c_2^{(1)}$ are not both zero, equation (6–8) would show that $w_1(z)$ and $w_2(z)$ are linearly dependent. But this is impossible since $w_1(z)$ and $w_2(z)$ are, by hypothesis, a fundamental set. Hence, we conclude that $\lambda_1 \neq 0$. The same reasoning shows that $\lambda_2 \neq 0$. We can, therefore, define two (finite) numbers δ_1 and δ_2 by [75]

$$\delta_1 = \frac{1}{2\pi i} \ln \lambda_1 \qquad \delta_2 = \frac{1}{2\pi i} \ln \lambda_2 \tag{6-12}$$

and use these numbers to define the two analytic functions $f_1(z)$ and $f_2(z)$ in D_0 by

$$\left. \begin{aligned} f_1(z) &= (z - z_0)^{-\delta_1} u_1(z) \\ f_2(z) &= (z - z_0)^{-\delta_2} u_2(z) \end{aligned} \right\} \tag{6-13}$$

Since the right-hand members of these equations are products of functions which can be analytically continued along any simple curve in Δ, it follows that $f_1(z)$ and $f_2(z)$ must also have this property. And since $w_1(z)$, $w_2(z)$, $(z - z_0)^{-\delta_1}$ and $(z - z_0)^{-\delta_2}$ have no singular points in Δ, it follows from equations (6–8), (6–9), and (6–13) that $f_1(z)$ and $f_2(z)$ have no singular points in this region.[76] Hence, these functions have no branch points in Δ. On the other hand, the analytic continuations $f_i(e^{2\pi i}(z - z_0) + z_0)$ of $f_i(z)$ for $i = 1, 2$ in a counterclockwise direction around any curve Γ which encloses z_0 are

$$f_i(e^{2\pi i}(z - z_0) + z_0) = (e^{2\pi i}(z - z_0))^{-\delta_i} u_i(e^{2\pi i}(z - z_0) + z_0) \qquad \text{for } i = 1, 2$$

---

[75] The condition $\lambda_1 \neq \lambda_2$ implies that $\delta_1 - \delta_2$ is not an integer.
[76] However, the point z_0 may be a singular point of these functions.

Hence, it follows from equations (6–10) and (6–11) to (6–13) that

$$f_i(e^{2\pi i}(z-z_0)+z_0) = \frac{1}{\lambda_i}\,(z-z_0)^{-\delta_i}u_i(e^{2\pi i}(z-z_0)+z_0)$$

$$= (z-z_0)^{-\delta_i}u_i(z) = f_i(z) \qquad \text{for } i=1, 2$$

But this shows that the point z_0 is also not a branch point of the functions $f_i(z)$, $i=1, 2$. Hence, these functions will return to their original values when they are analytically continued around any closed curve in Δ. And this implies, as indicated in section 5.7, that they can be extended to *single-valued* analytic functions on the *entire* domain Δ. We can therefore suppose that these extensions have already been carried out and that $f_1(z)$ and $f_2(z)$ are defined on the entire domain Δ. Hence, it follows from equations (6–13) that the extensions of the solutions $u_1(z)$ and $u_2(z)$ from D_0 to the domain Δ (which we shall also denote by $u_1(z)$ and $u_2(z)$) are functions of the form

$$\left.\begin{aligned} u_1(z) &= (z-z_0)^{\delta_1}f_1(z) \\[2mm] u_2(z) &= (z-z_0)^{\delta_2}f_2(z) \end{aligned}\right\} \tag{6-14}$$

where $f_1(z)$ and $f_2(z)$ are single-valued analytic functions on the entire domain Δ. (The point z_0, however, may be an isolated singular point of these functions.) And as indicated in section 5.8 the extended functions $u_1(z)$ and $u_2(z)$ must satisfy the differential equation (6–1) everywhere within Δ. These are the solutions which we have set out to construct. We shall now show that they are linearly independent. To this end suppose that γ_1 and γ_2 are any two constants such that

$$\gamma_1 u_1(z) + \gamma_2 u_2(z) = 0$$

Substituting equations (6–8) and (6–9) into this equation shows that

$$[\,\gamma_1 c_1^{(1)} + \gamma_2 c_1^{(2)}\,]\,w_1(z) + [\,\gamma_1 c_2^{(1)} + \gamma_2 c_2^{(2)}\,]\,w_2(z) = 0$$

And since $w_1(z)$ and $w_2(z)$ are linearly independent, it follows that

$$\left.\begin{array}{l} \gamma_1 c_1^{(1)} + \gamma_2 c_1^{(2)} = 0 \\[2mm] \gamma_1 c_2^{(1)} + \gamma_2 c_2^{(2)} = 0 \end{array}\right\} \tag{6-15}$$

Hence, upon multiplying equation (6–7a) by γ_1 and equation (6–7b) by γ_2 and adding the results, we find that

$$(\lambda_1 - \lambda_2)\gamma_1 c_1^{(1)} = (\lambda_1 - \lambda_2)\gamma_2 c_2^{(2)} = 0$$

But since we are considering the case where $\lambda_1 \neq \lambda_2$, this implies that

$$\gamma_1 c_1^{(1)} = \gamma_2 c_2^{(2)} = \gamma_2 c_1^{(2)} = \gamma_1 c_2^{(1)} = 0$$

And by using the facts that $c_1^{(1)}$ and $c_2^{(1)}$ are not both zero and $c_1^{(2)}$ and $c_2^{(2)}$ are not both zero, we can conclude that $\gamma_1 = \gamma_2 = 0$. But this shows that $u_1(z)$ and $u_2(z)$ are linearly independent.

Before discussing the implications of these solutions, we shall first treat the case where the roots of the characteristic equation are equal.

6.2.2 Case II: Roots of Characteristic Equation Equal

Thus, suppose that the roots of equation (6–6) are equal. Then the argument used in the preceding section can easily be adapted to show that there still exists at least one solution to equation (6–1) of the form

$$w_1(z) = (z - z_0)^{\delta_1} f_1(z) \tag{6-16}$$

where $f_1(z)$ is a single-valued analytic function in Δ and $\delta_1 = (1/2\pi i) \ln \lambda_1$ with $\lambda_1 \neq 0$. We can therefore assume that the fundamental set of solutions $w_1(z)$ and $w_2(z)$ of equation (6–1) in the domain D_0 (fig. 6–1) has been chosen in such a way that $w_1(z)$ is given by equation (6–16). By analytically continuing this solution around Γ we find that in this case the function $W_1(z)$ (defined in eqs. (6–3)) is given by $W_1(z) = \lambda_1 w_1(z)$. And since $w_1(z)$ and $w_2(z)$ are linearly independent, the first equation (6–4) shows that

$$a_{12} = 0$$

and

$$a_{11} = \lambda_1$$

However, these equations imply that the roots of the characteristic equation (6–6) are $\lambda_1 = a_{11} \neq 0$ and $\lambda_2 = a_{22}$. But since we are considering the case where $\lambda_1 = \lambda_2$, it follows that $a_{11} = a_{22} = \lambda_1$. Hence, in this case equations (6–4) become

$$\left.\begin{aligned}
W_1(z) &= \lambda_1 w_1(z) \\
W_2(z) &= a_{21} w_1(z) + \lambda_1 w_2(z)
\end{aligned}\right\} \qquad (6\text{–}17)$$

We now define the function $f_2(z)$ on the domain D_0 by

$$f_2(z) = (z - z_0)^{-\delta_1} w_2(z) - \frac{a_{21} f_1(z)}{2\pi i \lambda_1} \ln (z - z_0) \qquad (6\text{–}18)$$

This function can be analytically continued along any simple curve in Δ, and it will return to its original value when it is analytically continued along any closed curve which does not encircle z_0. Its analytic continuation (counterclockwise) along any curve Γ which encircles z_0 is

$$f_2(e^{2\pi i}(z - z_0) + z_0) = (z - z_0)^{-\delta_1} e^{-2\pi i \delta_1} W_2(z) - \frac{a_{21} f_1(z)}{2\pi i \lambda_1} \ln e^{2\pi i}(z - z_0)$$

$$= \frac{(z - z_0)^{-\delta_1}}{\lambda_1} W_2(z) - \frac{a_{21} f_1(z)}{2\pi i \lambda_1} \ln (z - z_0) - \frac{a_{21}}{\lambda_1} f_1(z)$$

But inserting equation (6–16) and the second equation (6–17) into this relation shows that

$$f_2(e^{2\pi i}(z - z_0) + z_0) = f_2(z)$$

Hence, the function $f_2(z)$ will return to its original value when it is analytically continued around *any* closed curve in Δ. It can therefore be extended to a single-valued analytic function on the entire domain (which may have an isolated singular point at z_0). And we can now conclude from equations (6–16) and (6–18) that the extension of the solution $w_2(z)$ from D_0 to the entire domain Δ (which we shall also denote by $w_2(z)$) is a function of the form

$$w_2(z) = (z - z_0)^{\delta_1} f_2(z) + aw_1(z) \ln (z - z_0) \tag{6–19}$$

where $f_2(z)$ is a single-valued analytic function on the entire domain Δ and we have put $a = a_{21}/2\pi i \lambda_1$. It can again be shown that the solutions $w_1(z)$ and $w_2(z)$ are linearly independent.

6.2.3 General Conclusions

Notice that the fundamental set of solutions (6–16) and (6–19) with $a = 0$ are of the same form as the fundamental set (6–14) with $\delta_1 = \delta_2$. The fundamental set (6–2) which occurs at an ordinary point of equation (6–1) is a special case of the fundamental set (6–16) and (6–19). We have therefore established the following conclusion: *Let z_0 be an ordinary point or an isolated singular point of equation (6–1) and let R be the distance between z_0 and the singular point of equation (6–1) which is closest to z_0. Then equation (6–1) possesses a fundamental set of solutions on the punctured circular region $0 < |z - z_0| < R$ which is of the form*

$$\left. \begin{aligned} w_1(z) &= (z - z_0)^{\delta_1} f_1(z) \\[2mm] w_2(z) &= (z - z_0)^{\delta_2} f_2(z) + aw_1(z) \ln (z - z_0) \end{aligned} \right\} \tag{6–20}$$

where $f_1(z)$ and $f_2(z)$ are analytic single-valued functions in $0 < |z - z_0| < R$ and a, δ_1, and δ_2 are complex constants.

The fundamental set (6–20) is called a *canonical basis*. The constant a is equal to zero whenever $\delta_1 - \delta_2 \neq 0, \pm 1, \pm 2, \ldots$. The case where $\delta_1 - \delta_2 = 0, \pm 1, \pm 2, \ldots$ is referred to as the *exceptional case*. Notice that the constant a may also vanish in the exceptional case. This occurs, for example, when z_0 is an ordinary point of equation (6–1).

We should not conclude from these results, however, that the solutions of equation (6–1) will always have singularities at the singular points of this equation. For example, the equation

$$z^2 w'' - 2zw' + 2w = 0$$

has an isolated singular point at $z = 0$. But the general solution to this equation is

$$w = A_0 z + A_1 z^2$$

where A_0 and A_1 are arbitrary. And this shows that every solution of this equation is analytic at $z = 0$.

By changing notation we can rewrite this solution in the form

$$w = a_0 + a_1(z - 1) + (a_1 - a_0)(z - 1)^2$$

where a_0 and a_1 can now be taken as arbitrary. This is evidently a canonical basis at the ordinary point $z = 1$. Since it is an entire function, it certainly exists and satisfies the equation within a circle whose radius is larger than the distance between $z = 1$ and the nearest singularity (namely, $z = 0$) of the differential equation.

If either of the functions $f_1(z)$ and $f_2(z)$ in equations (6–20) is not analytic at z_0, then z_0 must be an *isolated* singular point of this function. Hence, $f_1(z)$ and $f_2(z)$ can always be expanded in the Laurent series [77]

$$\left. \begin{aligned} f_1(z) &= \sum_{n=-\infty}^{\infty} b_n^{(1)}(z - z_0)^n \\[2mm] f_2(z) &= \sum_{n=-\infty}^{\infty} b_n^{(2)}(z - z_0)^n \end{aligned} \right\} \qquad (6\text{–}21)$$

which converge for $0 < |z - z_0| < R$. An important special case occurs when the functions $f_1(z)$ and $f_2(z)$ either have poles or are analytic at z_0. Then the series (6–21) contain at most a finite number of negative powers of $z - z_0$ and can therefore be written in the form

---

[77] Which may or may not contain negative powers of $(z - z_0)$.

$$f_i = \sum_{n=-r_i}^{\infty} b_n^{(i)}(z-z_0)^n \qquad \text{for } i=1, 2$$

where r_1 and r_2 are finite integers (they may be negative or zero). However, we can shift the index in the sums by putting $k=n+r_i$ and then summing on k. Thus,

$$f_i(z) = \sum_{k=0}^{\infty} b_{k-r_i}^{(i)}(z-z_0)^{k-r_i} = (z-z_0)^{-r_i} \sum_{k=0}^{\infty} a_k^{(i)}(z-z_0)^k \qquad \text{for } i=1, 2$$

where we have put $a_k^{(i)} \equiv b_{k-r_i}^{(i)}$ for $i=1, 2$.

Inserting these equations into equations (6–20) shows that in this case the canonical basis is of the form

$$\left.\begin{aligned}
w_1(z) &= (z-z_0)^{\rho_1} \sum_{k=0}^{\infty} a_k^{(1)}(z-z_0)^k \\[2ex]
w_2(z) &= (z-z_0)^{\rho_2} \sum_{k=0}^{\infty} a_k^{(2)}(z-z_0)^k + a w_1(z) \ln (z-z_0)
\end{aligned}\right\} \qquad (6\text{–}22)$$

where we have put $\rho_1 = \delta_1 - r_1$ and $\rho_2 = \delta_2 - r_2$. The series converge everywhere within the circle $|z-z_0| < R$ (including the point z_0) and they, therefore, represent analytic functions within this circle. The solutions w_1 and w_2 are said to be *regular*. It follows from the fact that r_1 and r_2 are integers that a vanishes whenever $\rho_1 - \rho_2 \neq 0, \pm 1, \pm 2, \ldots$ since it vanishes when $\delta_1 - \delta_2$ has this property.

We have already seen that equation (6–1) will always possess two regular solutions (with $\rho_1 = \rho_2 = a = 0$) at the point z_0 whenever z_0 is an ordinary point of this equation (i.e., if $p(z)$ and $q(z)$ are analytic at z_0). We shall now determine certain conditions which the coefficients of equation (6–1) must satisfy at an isolated singular point z_0 if this equation is to possess two regular solutions about z_0.

Since $w_1(z)$ and $w_2(z)$ are linearly independent in $0 < |z-z_0| < R$, their Wronskian (see section 5.9.3)

$$\mathcal{W}(z) = \begin{vmatrix} w_1(z) & w_2(z) \\ w_1'(z) & w_2'(z) \end{vmatrix} \qquad (6\text{–}23)$$

cannot vanish at any point z of this domain. On the other hand, putting $g_i \equiv \sum_{k=0}^{\infty} a_k^{(i)}(z - z_0)^k$ for $i = 1$, 2 in equations (6–22) and substituting the result into equation (6–23) shows that

$$\mathscr{W}(z) = (z - z_0)^{\rho_1 + \rho_2 - 1} G_1(z) + a(z - z_0)^{2\rho_1 - 1} G_2(z) \qquad (6\text{–}24)$$

where we have set $G_1 = (\rho_2 - \rho_1)g_1 g_2 + (z - z_0)(g_1 g_2' - g_1' g_2)$ and $G_2 \equiv g_1^2$. Since g_1 and g_2 are single-valued analytic functions in the entire circle $|z - z_0| < R$ (including z_0), it is clear that G_1 and G_2 are also functions of this type. And since $a = 0$ whenever the exponents $\rho_1 + \rho_2 - 1$ and $2\rho_1 - 1$ in equation (6–24) do not differ from one another by an integer, this equation can also be written as

$$\mathscr{W}(z) = (z - z_0)^{\lambda} G_0(z)$$

where λ is the smaller of the two numbers $\rho_1 + \rho_2 - 1$ and $2\rho_1 - 1$ and $G_0(z)$ is a single-valued analytic function in the entire circle $|z - z_0| < R$. Hence, $G_0(z)$ can be represented by a power series

$$G_0(z) = \sum_{n=0}^{\infty} A_n (z - z_0)^n$$

whose radius of convergence is R. But it follows from the fact that $\mathscr{W}(z)$ is not equal to zero in $0 < |z - z_0| < R$ that there must be a smallest integer, say m, such that $A_m \neq 0$. Hence,

$$\mathscr{W}(z) = (z - z_0)^{\lambda + m} G(z) \qquad (6\text{–}25)$$

where

$$G(z) = \sum_{k=0}^{\infty} A_{k+m}(z - z_0)^k = (z - z_0)^{-m} G_0(z)$$

is also a single-valued analytic function in $|z - z_0| < R$ but not equal to zero at z_0. And since $\mathscr{W}(z)$ does not vanish at any point of the punctured region

$0 < |z - z_0| < R$, we conclude from equation (6–25) that $G(z)$ is not equal to zero anywhere within the circle $|z - z_0| < R$ (including the point z_0). Hence, differentiating (6–25) shows that

$$\frac{1}{W} \frac{dW}{dz} = \frac{d \ln W}{dz} = \frac{\lambda + m}{z - z_0} - P(z)$$

where

$$P(z) \equiv -\frac{G'(z)}{G(z)}$$

is a single-valued analytic function in $|z - z_0| < R$ since its denominator does not vanish in this domain. But equation (5–26) shows that the coefficient $p(z)$ of w' in equation (6–1) is related to the Wronskian by

$$p(z) = -\frac{1}{W} \frac{dW}{dz}$$

Thus, if equation (6–1) possesses two regular solutions, its coefficient $p(z)$ must be of the form

$$p(z) = -\frac{\lambda + m}{z - z_0} + P(z) \tag{6–26}$$

Similarly, it can be shown by substituting the first equation (6–22) and equation (6–26) into equation (6–1) that this equation will possess two regular solutions at z_0 only if the coefficient $q(z)$ is of the form

$$q(z) = \frac{\alpha}{(z - z_0)^2} + \frac{\beta}{z - z_0} + Q(z)$$

where α and β are constants and $Q(z)$ is analytic at z_0. We have therefore shown that equation (6–1) can possess two regular solutions at the point z_0 only if its coefficient $p(z)$ has, at most, a simple pole and its coefficient $q(z)$ has at most a pole of order 2. We therefore say that a singular point z_0 of equation (6–1) is a *regular singular point* if

(1) $p(z)$ is either analytic at z_0 or has a simple pole at z_0

(2) $q(z)$ is either analytic at z_0, has a simple pole at z_0, or else has a pole of order 2 at z_0.

Notice, however, that we have not yet shown that equation (6–1) always possesses two regular solutions at any regular singular point.

An isolated singular point of equation (6–1) which is not a regular singular point is called an *irregular singular point*.

In order to treat the point $z = \infty$, we must, as in section 5.3, introduce the new independent variable $\zeta = 1/z$ into equation (6–1). Then the point $z = \infty$ will be an ordinary point, a regular singular point, or an irregular singular point if the point $\zeta = 0$ is, respectively, an ordinary point, a regular singular point, or an irregular singular point of the transformed equation. But the change of variable $\zeta = 1/z$ transforms equation (6–1) into the equation

$$\frac{d^2w}{d\zeta^2} + P_0(\zeta)\frac{dw}{d\zeta} + Q_0(\zeta)w = 0 \tag{6–27}$$

where

$$P_0(\zeta) \equiv \frac{2}{\zeta} - \frac{1}{\zeta^2} p\left(\frac{1}{\zeta}\right) = 2z - z^2 p(z) \tag{6–28}$$

$$Q_0(\zeta) \equiv \frac{1}{\zeta^4} q\left(\frac{1}{\zeta}\right) = z^4 q(z) \tag{6–29}$$

Thus $z = \infty$ is an ordinary point of equation (6–1) if both $P_0(\zeta)$ and $Q_0(\zeta)$ are analytic at $\zeta = 0$, which is equivalent to saying that $2z - z^2 p(z)$ and $z^4 q(z)$ are analytic at $z = \infty$. Similarly, the point $z = \infty$ is a regular singular point of equation (6–1) if $P_0(\zeta)$ has, at most, a simple pole at $\zeta = 0$ and $Q_0(\zeta)$ has, at most, a pole of order 2 at $\zeta = 0$, which is equivalent to saying that $2z - z^2 p(z)$ has, at most, a simple pole at $z = \infty$ and $z^4 q(z)$ has, at most, a pole of order 2 at $z = \infty$. If the point $z = \infty$ is an isolated singular point of equation (6–1) but is not a regular singular point, it is an irregular singular point. It is clear that the coefficients $p(z)$ and $q(z)$ themselves must certainly be analytic at $z = \infty$ whenever this point is a regular singular point.

These definitions are best clarified by considering an example. Thus, the differential equation

$$z^2(z-1)w'' - (z+1)w' + zw = 0$$

can be written as

$$w'' - \frac{(z+1)}{z^2(z-1)} w' + \frac{1}{z(z-1)} w = 0$$

Hence,

$$p(z) = -\frac{(z+1)}{z^2(z-1)} \qquad q(z) = \frac{1}{z(z-1)}$$

And the function p has a simple pole at $z=1$ and a pole of order 2 at the point $z=0$. The point $z=0$ must, therefore, be an irregular singular point. The function q has simple poles at $z=0$ and $z=1$. Hence, the point $z=1$ is a regular singular point. All other finite points are ordinary points. In order to consider the point at infinity we must consider the functions $2z - z^2 p(z)$ and $z^4 q(z)$ (instead of the coefficients p and q) which, in this case, become

$$2z - z^2 p = 2z + \frac{z+1}{z-1} \qquad z^4 q = \frac{z^3}{z-1}$$

Hence, $2z - z^2 p$ has a simple pole and $z^4 q$ has a pole of order 2. And this shows that the point at ∞ is a regular singular point.

We shall now show by actually giving a procedure for finding the solutions that the differential equation (6–1) always possesses two regular solutions at a regular singular point.

6.3 SOLUTION OF EQUATION ABOUT ORDINARY POINTS AND REGULAR SINGULAR POINTS

Let z_0 be either an ordinary point or a regular singular point of the differential equation

$$w'' + p(z)w' + q(z)w = 0 \tag{6-30}$$

Since p has, at most, a simple pole and q has, at most, a pole of order 2 at z_0, the Laurent series expansion for p about z_0 contains, at most, one negative power of $z - z_0$ and that for q contains, at most, two. Let R_1 and R_2 be the distances between z_0 and the nearest singular points of $p(z)$ and $q(z)$, respectively, to z_0. Then R_1 and R_2 will be strictly positive and

$$p(z) = \sum_{k=0}^{\infty} p_k (z - z_0)^{k-1} \qquad \text{for } 0 < |z - z_0| < R_1 \qquad (6\text{–}31)$$

$$q(z) = \sum_{k=0}^{\infty} q_k (z - z_0)^{k-2} \qquad \text{for } 0 < |z - z_0| < R_2 \qquad (6\text{–}32)$$

The point z_0 will be an ordinary point of the differential equation if, and only if,

$$p_0 = q_0 = q_1 = 0 \qquad (6\text{–}33)$$

6.3.1 First Regular Solution

We shall now show that equation (6–30) always has at least one solution of the form

$$w(z) = (z - z_0)^{\rho} \sum_{n=0}^{\infty} a_n (z - z_0)^n \qquad (6\text{–}34)$$

in some punctured circular region about the point z_0. This will be accomplished by first showing that the constants ρ and a_n can always be determined in such a way that equation (6–34) formally satisfies the differential equation. It will then be shown, a posteriori, that the formal operations were indeed justified and therefore that the formal solution is, in fact, a true solution of the differential equation. The procedure which we develop by this process can then always be used to construct a solution to *any* given differential equation of the form (6–30) in the neighborhood of any of its ordinary points or regular singular points.

Since the assumed expansion (6–34) must certainly have a leading term and since the exponent ρ is not, as yet, specified, we can always assume that matters are arranged so that $a_0 \neq 0$. We now substitute this expansion into the differential equation (6–30) and suppose that the series can be differentiated term by term. Then

$$\sum_{n=0}^{\infty} (n+\rho)(n+\rho-1)a_n(z-z_0)^{n+\rho-2}$$

$$+ \left[\sum_{k=0}^{\infty} p_k(z-z_0)^{k-1} \right] \left[\sum_{n=0}^{\infty} (n+\rho)a_n(z-z_0)^{n+\rho-1} \right]$$

$$+ \left[\sum_{k=0}^{\infty} q_k(z-z_0)^{k-2} \right] \left[\sum_{n=0}^{\infty} a_n(z-z_0)^{n+\rho} \right] = 0$$

And after forming the Cauchy products of the series and adding the resulting series term by term, we get

$$\sum_{n=0}^{\infty} (z-z_0)^{n+\rho-2} \left\{ (n+\rho)(n+\rho-1)a_n + \sum_{k=0}^{n} a_k \left[(k+\rho)p_{n-k} + q_{n-k} \right] \right\} = 0$$

$$(6\text{--}35)$$

Now it follows from the uniqueness property of power series that this series will vanish for all values of z in some domain only if the coefficients of each power of $z - z_0$ vanish individually. Upon equating these coefficients to zero, we obtain for $n=0$

$$F(\rho)a_0 = 0 \qquad\qquad (6\text{--}36)$$

and

$$F(n+\rho)a_n = - \sum_{k=0}^{n-1} a_k[(k+\rho)p_{n-k} + q_{n-k}] \qquad \text{for } n = 1, 2, \ldots$$

$$(6\text{--}37)$$

where we have put

$$F(\rho) \equiv \rho^2 + (p_0 - 1)\rho + q_0 \qquad\qquad (6\text{--}38)$$

Since, by hypothesis, $a_0 \neq 0$, equation (6–36) implies

$$F(\rho) \equiv \rho^2 + (p_0 - 1)\rho + q_0 = 0 \qquad (6\text{-}39)$$

This quadratic equation is called the *indicial equation*. Its two roots are called the *characteristic exponents* of the differential equation at the point z_0. Equation (6–37) is called the *recurrence relation*.

If the constant ρ in the assumed solution (6–34) is chosen to be a root of the indicial equation (6–38), the coefficient of $(z - z_0)^{\rho-2}$ in equation (6–35) will vanish for arbitrary values of the constant a_0. And if *there is no positive integer n for which $F(n + \rho) = 0$*, equation (6–37) can be used to calculate successively (starting with a_0) the coefficients [78] a_n in such a way that the coefficients of all the remaining powers of $z - z_0$ in equation (6–35) will vanish. Thus, when the constants ρ and a_n for $n = 1, 2, \ldots$ are determined in this manner, the series (6–34) will at least formally satisfy the differential equation (6–30) with the constant a_0 arbitrary.

Of course, if $F(n + \rho)$ vanishes for some positive integer n, we have no assurance that equation (6–37) can be used to calculate a_n. But let the roots of the indicial equation (i.e., the characteristic exponents) be denoted by ρ_1 and ρ_2 with the notation chosen so that $\mathscr{Re}\ \rho_1 \geqslant \mathscr{Re}\ \rho_2$, and let

$$\nu \equiv \rho_1 - \rho_2 \qquad (6\text{-}40)$$

Then

$$\mathscr{Re}\ \nu \geqslant 0 \qquad (6\text{-}41)$$

And since a quadratic function can always be expressed as the product of its factors, we can write

$$F(\rho) = (\rho - \rho_1)(\rho - \rho_2)$$

Hence,

$$F(n + \rho_1) = n(n + \nu) \qquad (6\text{-}42)$$

---

[78] Notice that for each integer n the recurrence relation (6–37) expresses a_n only in terms of the coefficients a_k with $0 \leqslant k < n$. Hence, it can be used to first determine a_1 in terms of a_0 and then to determine a_2 in terms of a_1 and a_0. But since a_1 is known in terms of a_0, this determines a_2 in terms of a_0. By proceeding in this manner, each a_n can be determined in succession in terms of a_0.

Then, since equation (6–41) shows that

$$\mathscr{R}_e \, (n+\nu) = n + \mathscr{R}_e \, \nu \geq n \tag{6-43}$$

it follows from equation (6–42) that

$$F(n+\rho_1) \neq 0 \qquad \text{for } n=1, 2, \ldots \tag{6-44}$$

Hence, when $\rho = \rho_1$, the coefficients a_n for $n=1, 2, \ldots$ can all be calculated recursively in terms of a_0 from the recurrence relation (6–37). Thus, equation (6–30) will always possess at least one formal solution of the form (6–34) about the point z_0.

It is still necessary to verify that this formal solution is justified. In order to do this, we must show that the series $\sum_{n=0}^{\infty} a_n(z-z_0)^n$ obtained by the procedure described actually converges. We can then conclude from the theory of power series given in section 5.5 that the formal operations of (1) differentiating the series (6–34) term by term, (2) forming the Cauchy product of the resulting convergent series with the convergent series (6–31) and (6–32), and (3) adding the resulting convergent series term by term are justified [79] and, hence, that equation (6–34) is indeed a solution to the differential equation (6–30).

In order to establish the convergence of the series in equation (6–34) for the case where $\rho = \rho_1$, notice that in view of equations (6–42) and (6–44) the recurrence relation (6–37), with $\rho = \rho_1$, can be put in the form

$$a_n = -\frac{\sum_{k=0}^{n-1} a_k[(k+\rho_1)p_{n-k} + q_{n-k}]}{n(n+\nu)} \qquad \text{for } n = 1, 2, \ldots \tag{6-45}$$

---

[79] We shall subsequently encounter (in ch. 9) a case where the constructed series actually diverges and hence the formal procedures are not necessarily justified.

Therefore,

$$|a_n| = \frac{\left| \sum\limits_{k=0}^{n-1} a_k [(k+\rho_1)p_{n-k} + q_{n-k}] \right|}{|n(n+\nu)|}$$

$$\leq \frac{\sum\limits_{k=0}^{n-1} |a_k| [|p_{n-k}||\rho_1| + |q_{n-k}| + k|p_{n-k}|]}{n|n+\nu|} \qquad (6\text{--}46)$$

Let $R = \min\{R_1, R_2\}$, where R_1 and R_2 are the radii of convergence of the series (6–31) and (6–32) for p and q, respectively. Then R is *strictly positive* (since R_1 and R_2 are positive) and the Cauchy estimates, given in section 5.5, for the derivatives of the analytic functions p and q imply that there exist *finite* positive constants M and N such that

$$|p_k| \leq \frac{M}{R^k} \qquad \text{and} \qquad |q_k| \leq \frac{N}{R^k} \qquad \text{for } k = 0, 1, 2, \ldots \qquad (6\text{--}47)$$

which implies that

$$|p_k||\rho_1| + |q_k| \leq \frac{M|\rho_1| + N}{R^k} \qquad \text{for } k \geq 1$$

And since $|n+\nu| \geq \mathcal{R}e(n+\nu)$, it follows from equation (6–43) that $|n+\nu| \geq n$ for $n = 1, 2, \ldots$. Inserting these results into equation (6–46) shows that

$$|a_n| \leq \frac{1}{n} \sum_{k=0}^{n-1} \left(\frac{M|\rho_1| + N}{n} + \frac{kM}{n} \right) \frac{|a_k|}{R^{n-k}} \qquad \text{for } n = 1, 2, \ldots \qquad (6\text{--}48)$$

Now put $P \equiv M|\rho_1| + N + M + 1$. Then P is a *finite* number which is larger than 1; and for $n \geq 1$ and $k < n$,

$$\frac{M|\rho_1|+N+kM}{n} \leqslant \frac{M|\rho_1|+N+M}{n} \leqslant P$$

Hence, it follows from equation (6–48) that

$$|a_n| \leqslant \frac{P}{n} \sum_{k=0}^{n-1} \frac{|a_k|}{R^{n-k}} \qquad \text{for } n=1,2,\ldots \qquad (6\text{–}49)$$

which for $n=1$ becomes

$$|a_1| \leqslant \left(\frac{P}{R}\right)^1 |a_0| \qquad (6\text{–}50)$$

Suppose that for $n > 1$

$$|a_k| \leqslant \left(\frac{P}{R}\right)^k |a_0| \qquad \text{for } 1 \leqslant k < n \qquad (6\text{–}51)$$

We shall now show that this implies that the inequality also holds when $k = n$. Since n can be any integer larger than 1 and since equation (6–50) shows that the inequality also holds when $n=1$, we can then conclude by induction that the inequality will hold for every positive integer n. Thus, inserting the inequality (6–51) into equation (6–49) shows that

$$|a_n| \leqslant \frac{P}{n} \sum_{k=0}^{n-1} \frac{P^k}{R^n} |a_0| \qquad (6\text{–}52)$$

But since $P \geqslant 1$, it follows that $P^k \leqslant P^{n-1}$. And when this is inserted into equation (6–52), we find that

$$|a_n| \leqslant \left(\frac{P}{R}\right)^n |a_0| \qquad (6\text{–}53)$$

Hence, we can conclude by induction that this inequality holds for every positive integer n.

We can now use this inequality to show that the series $\sum\limits_{n=0}^{\infty} a_n(z-z_0)^n$, which appears in the solution (6–34), converges at least within the circular region (of nonzero radius)

$$|z - z_0| \leq \frac{R}{2P} \tag{6-54}$$

To this end, notice that within this circle the inequality (6–53) shows that the nth term of this series has the property that

$$|a_n(z - z_0)^n| = |a_n| \, |z - z_0|^n \leq \left(\frac{P}{R}\right)^n |a_0| \left(\frac{R}{2P}\right)^n = |a_0| \frac{1}{2^n} \tag{6-55}$$

And since the series

$$|a_0| \sum_{n=0}^{\infty} \frac{1}{2^n}$$

is simply a geometric series, it certainly converges. Hence, in view of the inequality (6–55), a simple application of the comparison test [80] shows that the series $\sum_{n=0}^{\infty} a_n(z - z_0)^n$ is absolutely and uniformly convergent at least within the circle (6–54). It, therefore, represents an (single valued) analytic function within this circle.

Thus, we have shown that

$$w_1(z) = (z - z_0)^{\rho_1} \sum_{n=0}^{\infty} a_n^{(1)} (z - z_0)^n \tag{6-56}$$

is a solution to equation (6–30) within some circle about z_0 for arbitrary $a_0^{(1)}$ provided the coefficients $a_n^{(1)}$ for $n = 1, 2, \ldots$ are computed from the recurrence relation (6–45). But the series in (6–56) converges within a circle about z_0 which passes through the singular point of $w_1(z)$ nearest to z_0. In fact, the radius of convergence of this series must be at least as large as the distance R between z_0 and the nearest singular point of equation (6–30). In order to prove this, notice that R_1 and R_2, the radii of convergence of the series for p and q, respectively, must be equal to the distance between z_0 and the nearest singular

---

[80] See ref. 23.

points of these functions. But it follows from the definition of a singular point of a differential equation that $R = \min \{R_1, R_2\}$. Hence, p and q are single valued and analytic within the punctured circular region Δ defined by $0 < |z - z_0| < R$. Now since w_1 is a solution to the differential equation, it must be possible (as shown in section 5.9.2) to analytically continue this function along any simple curve in Δ. And, therefore, w_1 cannot have any singular points within Δ. But it certainly has a singular point on its circle of convergence. Hence, its radius of convergence cannot be smaller than R. And this proves the assertion.

Since the analytic continuation of a solution of the differential equation (6–30) is also a solution of this equation, we can now conclude that the *function* $w_1(z)$ *given by equation* (6–56) *converges and satisfies the differential equation* (6–30) *at every point of the domain* $0 < |z - z_0| < R$, *where* R *is the distance between* z_0 *and the nearest singular point of equation* (6–30).

We have now shown that equation (6–30) possesses a regular solution (6–56) about any ordinary point or regular singular point of this equation.

Before obtaining a second regular solution we shall introduce an example to illustrate the procedure described in the preceding paragraphs. Thus, consider the equation

$$3z^2 w'' + zw' - (1 + z)w = 0 \tag{6–57}$$

Since the coefficients p and q are

$$p(z) = \frac{1}{3z} \qquad q = -\frac{1+z}{3z^2}$$

and since the coefficients (6–28) and (6–29) of the transformed equation are

$$2z - z^2 p = \frac{5z}{3} \qquad z^4 q = -\frac{z^2(1+z)}{3}$$

we see that every point is an ordinary point except the point $z = 0$, which is a regular singular point, and the point $z = \infty$, which is an irregular singular point.

We know that if we seek a solution about the regular singular point $z = 0$ of the form

$$w = \sum_{n=0}^{\infty} a_n z^{n+\rho} \tag{6-58}$$

we will obtain at least one and perhaps two solutions to the differential equation in which the constant a_0 is arbitrary. And since the nearest singular point of the equation to $z = 0$ is at ∞, we know that the series will converge in the entire plane.

Instead of using the general formulas obtained above it is usually easier in any given case to derive the solution from first principles by substituting the assumed power series into the differential equation. Thus, after substituting equation (6–58) into equation (6–57) and differentiating term by term, we get

$$\sum_{n=0}^{\infty} [3(n+\rho)^2 - 2(n+\rho) - 1] a_n z^{n+\rho} - \sum_{n=0}^{\infty} a_n z^{n+\rho+1} = 0$$

In order to collect the coefficients of like powers of z, it is convenient to reindex the sums so that the same powers of z appear in each term. Thus, in the second sum, we replace the dummy index n by the index $k = n + 1$ and in the first sum we put $n = k$. And since $k = 1$ when $n = 0$, in the second sum we obtain

$$\sum_{k=0}^{\infty} [3(k+\rho)^2 - 2(k+\rho) - 1] a_k z^{k+\rho} - \sum_{k=1}^{\infty} a_{k-1} z^{k+\rho} = 0$$

Now equating to zero the coefficients of like powers of z gives

$$[3\rho^2 - 2\rho - 1] a_0 = 0 \qquad \text{for } k = 0$$

and

$$[3(k+\rho)^2 - 2(k+\rho) - 1] a_k = a_{k-1} \qquad \text{for } k = 1, 2, \ldots \tag{6-59}$$

Since $a_0 \neq 0$, the indicial equation is

$$F(\rho) = 3\rho^2 - 2\rho - 1$$

And its roots are

$$\rho_1 = 1 \qquad \text{and} \qquad \rho_2 = 1/3 \qquad\qquad (6\text{--}60)$$

When $\rho = 1$, the recurrence relation (6–59) can be solved for all values of k and can be written as

$$a_k = \frac{a_{k-1}}{k(3k+4)} \qquad \text{for } k = 1, 2, \ldots$$

or writing out the first few terms

$$a_1 = \frac{a_0}{1 \times 7}$$

$$a_2 = \frac{a_1}{2 \times 10}$$

$$\cdot$$
$$\cdot$$
$$\cdot$$

$$a_n = \frac{a_{n-1}}{n(3n+4)}$$

In order to determine a_n in terms of a_0 we multiply together both members of these equations to obtain

$$a_1 \cdot a_2 \ldots a_n = \frac{1}{(1 \cdot 2 \ldots n)\,[7 \cdot 10 \ldots (3n+4)]} \, a_0 \cdot a_1 \ldots a_{n-1}$$

and then divide through by $a_1 \cdot a_2 \ldots a_{n-1}$ to get

$$a_n = \frac{a_0}{n!\,[7 \cdot 10 \ldots (3n+4)]} \qquad \text{for } n = 1, 2, \ldots \qquad (6\text{--}61)$$

Notice that the term in square brackets is a product whose factors are elements of an arithmetic progression. The factors in this progression increase by 3 in each term. Hence, if we factor out a 3 from each term we obtain

$$7 \cdot 10 \cdot 13 \ldots (3n + 4) = 3^n \left(\frac{7}{3}\right)\left(\frac{7}{3} + 1\right)\left(\frac{7}{3} + 2\right) \ldots \left(\frac{7}{3} + n - 1\right) = 3^n \left(\frac{7}{3}\right)_n$$

where we have the generalized factorial function defined in equation (5–33). Upon substituting this into equation (6–61) and then substituting the result together with $\rho = 1$ into equation (6–58), we obtain the regular solution

$$w_1 = A_0 \sum_{n=0}^{\infty} \frac{1}{n! \left(\frac{7}{3}\right)_n} \left(\frac{z}{3}\right)^{n+1} \tag{6–62}$$

where we have put $A_0 = 3a_0$.

6.3.2 Second Regular Solution

We shall now use the regular solution $w_1(z)$ obtained in section 6.3.1 to show that equation (6–30) always possesses a second regular solution about any ordinary point or regular singular point z_0. In order to do this we use the results obtained in section 5.9.3, which show that

$$w_2(z) = w_1(z) \int \frac{dz}{[w_1(z)]^2} e^{-\int \rho(z)\,dz} \tag{6–63}$$

is a solution of equation (6–30) which is linearly independent of $w_1(z)$. We have omitted the arbitrary constant c in equation (5–28) since we may assume that it has been absorbed into the arbitrary constant a_0 which multiplies $w_1(z)$.

Now upon integrating equation (6–31) along any path in the punctured circular region $0 < |z - z_0| < R_1$, we obtain

$$\int p(z)\,dz = p_0 \ln(z - z_0) + c_0 + g_0(z) \tag{6–64}$$

where

$$g_0(z) \equiv \sum_{k=1}^{\infty} \frac{p_k}{k} (z - z_0)^k \tag{6–65}$$

and c_0 is an arbitrary constant.

185

Since the series $\sum_{k=1}^{\infty} p_k(z-z_0)^k$ converges, the series (6–65) must also converge (section 5.6). Hence, $g_0(z)$ is an analytic function in a circle centered at z_0. It, therefore, follows from equation (6–56) that

$$\frac{e^{-\int^z p(z)dz}}{[w_1(z)]^2} = (z-z_0)^{-2\rho_1}e^{-\rho_0 \ln (z-z_0)}g(z)$$

$$= (z-z_0)^{-(2\rho_1+p_0)}g(z) \qquad (6\text{–}66)$$

where we have put

$$g(z) \equiv e^{-c_0}\frac{e^{-g_0(z)}}{[g_1(z)]^2} \qquad \text{and} \qquad g_1(z) \equiv \sum_{n=0}^{\infty} a_n^{(1)}(z-z_0)^n$$

Since $a_0^{(1)} \neq 0$, there is some neighborhood of the point z_0 in which $g_1(z)$ is never equal to zero And since $g_1(z)$ is analytic at z_0, it follows that the function $[g_1(z)]^{-2}$ is also analytic and not equal to zero [81] at z_0. Similarly, it follows from the fact that $g_0(z)$ is analytic at z_0 that $\exp[-g_0(z)]$ is analytic and nonzero [81] at z_0. Hence, the function $g(z)$ is analytic and not equal to zero at z_0. It can, therefore, be expanded in the power series

$$g(z) = \sum_{n=0}^{\infty} \alpha_n(z-z_0)^n \qquad \text{with } \alpha_0 \neq 0 \qquad (6\text{–}67)$$

which converges in a circle of finite radius about z_0.

Now since the two roots ρ_1 and ρ_2 of the quadratic indicial equation (6–39) are given by

$$\frac{-(p_0-1) \pm \sqrt{(p_0-1)^2-4q_0}}{2}$$

it follows that $\rho_1+\rho_2=1-p_0$. And therefore, in view of definition (6–40),

$$2\rho_1+p_0=\nu+1 \qquad (6\text{–}68)$$

---

[81] Notice that $[g_1(z)]^{-2}$ and $\exp[-g_0(z)]$ can only equal zero at points where $g_1(z)$ and $g_0(z)$, respectively, become infinite; i.e., at singular points of these functions.

When equations (6–67) and (6–68) are substituted in equation (6–66) and the resulting expression is substituted into the integrand of equation (6–63), we find that

$$w_2(z) = w_1(z) \int \frac{1}{(z-z_0)^{\nu+1}} \sum_{n=0}^{\infty} \alpha_n (z-z_0)^n dz$$

And it follows from the fact that the power series converges that we can interchange the order of summation and integration (section 5.6) to obtain

$$w_2(z) = \begin{cases} w_1(z) \sum_{n=0}^{\infty} \frac{\alpha_n}{n-\nu}(z-z_0)^{n-\nu} & \text{for } \nu \neq 0, 1, 2, \ldots \\ w_1(z) \sum_{\substack{n=0 \\ n \neq \nu}}^{\infty} \frac{\alpha_n}{n-\nu}(z-z_0)^{n-\nu} + w_1(z)\alpha_\nu \ln (z-z_0) & \text{for } \nu = 0, 1, 2, \ldots \end{cases} \quad (6-69)$$

Since these series converge, they represent analytic functions within their circle of convergence. Therefore, the product of either of these series with the series in equation (6–56) is again an analytic function which can be expanded in a power series $\sum_{n=0}^{\infty} a_n^{(2)} (z-z_0)^n$ with a nonzero radius of convergence. Hence, in view of definition (6–40), it follows from equation (6–56) that equation (6–69) can be put in the form

$$w_2(z) = (z-z_0)^{\rho_2} \sum_{n=0}^{\infty} a_n^{(2)}(z-z_0)^n + aw_1(z) \ln (z-z_0) \quad (6-70)$$

where $a = 0$ for $\nu \neq 0, 1, 2, \ldots$

By repeating the argument used in section 6.3.1 to determine the size of the circle of convergence of the solution (6- 56), we can again show that *the function $w_2(z)$ given by equation (6–70) converges and satisfies the differential equation (6–30) at every point of the domain $0 < |z-z_0| < R$, where R is the distance between z_0 and the nearest singular point of equation (6–30).*

We have now proved that *the linear second-order differential equation (6–30) possesses two linearly independent regular solutions at every ordinary point and at every regular singular point.*

In order to show how to construct the solution (6–70), it is necessary to consider certain cases separately.

6.3.2.1 *Case (i):* $\nu \neq 0, 1, 2, \ldots$ —First, consider the case where the difference in the characteristic exponents ν is not equal to an integer. Then $a = 0$ in equation (6–70), and this solution is of the same form as the trial solution (6–34) with $\rho = \rho_2$. Hence, if equation (6–70) is substituted into the differential equation (6–30), we will find, as before (since ρ_2 already satisfies the indicial equation), that the coefficient $a_n^{(2)}$ must satisfy the recurrence relation (6–37) with $\rho = \rho_2$. But since equation (6–40) and the equation following equation (6–41) show that $F(n + \rho_2) = n(n - \nu)$, the recurrence relation (6–37) with $\rho = \rho_2$ can be written in the form

$$n(n-\nu)a_n = -\sum_{k=0}^{n-1} a_k[(k+\rho_2)p_{n-k} + q_{n-k}] \qquad \text{for } n = 1, 2, \ldots \qquad (6\text{--}71)$$

And since ν is not equal to an integer, the coefficient of a_n never vanishes. Hence, this equation can be solved recursively to determine the coefficients $a_n^{(2)}$ in exactly the same way that the coefficients $a_n^{(1)}$ in the first solution (6–56) were determined. Nothing new is involved. In fact, in solving the recurrence relation in this case it is usually easier to leave ρ unspecified as long as possible and to determine the coefficients of the first and second solutions simultaneously.

For example, the left member of the recurrence relation (6–59) of the differential equation (6–57) considered in section 6.3.1 can be factored to obtain

$$a_k = \frac{a_{k-1}}{(\rho + k - 1)(3\rho + 3k + 1)} \qquad \text{for } k = 1, 2, \ldots$$

or, writing out the first few terms,

$$a_1 = \frac{a_0}{\rho(3\rho+4)}$$

$$a_2 = \frac{a_1}{(\rho+1)(3\rho+7)}$$

.

.

.

$$a_n = \frac{a_{n-1}}{(\rho+n-1)(3\rho+3n+1)}$$

And upon multiplying together both members of these equations we obtain

$$a_1 \cdot a_2 \cdot a_3 \ldots a_n$$

$$= \frac{a_0 \cdot a_1 \cdot a_2 \ldots a_{n-1}}{[\rho(\rho+1) \ldots (\rho+n-1)][(3\rho+4)(3\rho+7) \ldots (3\rho+3n+1)]}$$

Hence,

$$a_n = \frac{a_0}{[\rho(\rho+1) \ldots (\rho+n-1)][(3\rho+4)(3\rho+7) \ldots (3\rho+3n+1)]}$$

The characteristic exponents ρ_1 and ρ_2 are given by equation (6–60). If we set $\rho = \rho_1 = 1$ in this equation, we obtain the coefficients of the first solution. But if we set $\rho = \rho_2$, we obtain the coefficients

$$a_n = \frac{a_0}{\left[\left(\frac{-1}{3}\right)\cdot\left(\frac{2}{3}\right)\cdot\left(\frac{5}{3}\right)\cdots\left(\frac{3n-4}{3}\right)\right][3\cdot6\cdot9\ldots3n]}$$

$$= \frac{a_0}{[(-1)\cdot2\cdot5\ldots(3n-4)]n!}$$

of the second solution. After using the generalized factorial function (eq. (5–27)) to get

$$(-1)\cdot2\cdot5\ldots(3n-4) = 3^n\left(-\frac{1}{3}\right)\left(-\frac{1}{3}+1\right)\left(-\frac{1}{3}+2\right)\cdots\left(-\frac{1}{3}+n-1\right)$$

$$= 3^n\left(-\frac{1}{3}\right)_n$$

and then substituting the results with $\rho=1/3$ into equation (6–58), we find that the second solution to equation (6–57) is

$$w_2 = B_0 \sum_{n=0}^{\infty} \frac{1}{n!\left(-\frac{1}{3}\right)_n}\left(\frac{z}{3}\right)^{n-(1/3)}$$

where we have put $B_0 = b_0 3^{-1/3}$.

6.3.2.2 *Case (ii): $\nu = 1, 2, \ldots$ and a $= 0$.* — In this case the second solution (6–70) will also be of the same form as the trial solution (6–34) with $\rho=\rho_2$. Hence, its coefficients $a_n^{(2)}$ for $n=1, 2, \ldots$ must again be determined by the recurrence relation (6–71) with $\rho=\rho_2$. But since ν is a positive integer, the left-hand side of this equation will vanish when $n=\nu$. The recurrence relation for a_ν can therefore not be satisfied unless the right-hand side of this equation

$$\sum_{k=0}^{\nu=1} a_k[(k+\rho_2)p_{\nu-k}+q_{\nu-k}]$$

also vanishes, in which case it will be satisfied automatically for all values of a_ν. We then obtain a solution involving two arbitrary constants. And since this solution must be a general solution to the differential equation, it will contain the solution obtained for $\rho=\rho_1$.

This situation will always occur in the important special case when z_0 is an ordinary point of the differential equation. For in this case, equations (6–33) and (6–39) show that the indicial equation is

$$F(\rho) \equiv \rho(\rho - 1) = 0$$

Hence, the characteristic exponents are $\rho_1 = 1$ and $\rho_2 = 0$; and their difference $\nu \equiv \rho_1 - \rho_2 = 1$ is a positive integer.[82] But it follows from equation (6–33) that the recurrence relation (6–71) for a_1 becomes

$$1 \times 0 \times a_1 = a_0 \left[(0 \times p_1) + 0 \right]$$

and this is automatically satisfied for any choice of a_1. The recurrence relation (6–71) with $n = 2, 3, \ldots$ will then uniquely determine the remaining a_n. We therefore obtain in this case a solution which involves two arbitrary constants.

For example, consider the equation

$$(z^2 - 1) w'' + 6zw' + 4w = 0 \tag{6–72}$$

Since its coefficients p and q are

$$p(z) = \frac{6z}{z^2 - 1} \qquad q(z) = \frac{4}{z^2 - 1}$$

we see that the only singular points of this equation are regular singular points at $z = 1$, $z = -1$, and $z = \infty$.

We know that if we seek a solution about the ordinary point $z = 0$ of the form

$$w = \sum_{n=0}^{\infty} a_n z^n \tag{6–73}$$

we will obtain a general solution to the differential equation in which the coefficients a_0 and a_1 are arbitrary constants. And the series will converge at least within the circle $|z| < 1$.

---

[82] That ν must be an integer in this case could easily be anticipated from the fact that the characteristic exponents themselves must be integers if the solutions are to be analytic at $z = z_0$.

Thus, after substituting the expansion (6–73) into equation (6–72), differentiating term by term, collecting terms, and factoring the coefficient of z^n, we obtain

$$-\sum_{n=0}^{\infty} n(n-1)a_n z^{n-2} + \sum_{n=0}^{\infty} (n+4)(n+1)a_n z^n = 0$$

which becomes, upon shifting the index in the second sum,

$$-\sum_{k=0}^{\infty} k(k-1)a_k z^{k-2} + \sum_{k=2}^{\infty} (k+2)(k-1)a_{k-2} z^{k-2} = 0$$

We now equate to zero the coefficients of like powers of z to get

$$0 \times (-1)a_0 = 0 \qquad\qquad 1 \times 0 \times a_1 = 0$$

and

$$-k(k-1)a_k + (k+2)(k-1)a_{k-2} = 0 \qquad \text{for } k=2, 3, \ldots$$

The first two equations merely serve to show that a_0 and a_1 are arbitrary, as we already know. And since the coefficient of a_k is not zero for $k \geq 2$, we can write the remaining equations as

$$a_k = -\frac{k+2}{k} a_{k-2} \qquad \text{for } k=2, 3, \ldots \qquad (6\text{--}74)$$

This equation shows that each succeeding a_k is determined from the a_j whose subscript is 2 lower than its own. Thus, this recurrence relation will ultimately determine the a_k with even values of k in terms of a_0 and the a_k with odd values of k in terms of a_1. It is, therefore, convenient to consider separately the equations for even and odd values of k. First, upon writing out these equations for even values of k beginning with a_2, we find that

$$a_2 = -\frac{4}{2} \, a_0$$

$$a_4 = -\frac{6}{4} \, a_2$$

$$.$$
$$.$$
$$.$$

$$a_{2j} = -\frac{2(j+1)}{2j} \, a_{2j-2}$$

Similarly, for odd values of k beginning with a_3, we find that

$$a_3 = -\frac{5}{3} \, a_1$$

$$a_5 = -\frac{7}{5} \, a_3$$

$$.$$
$$.$$
$$.$$

$$a_{2j+1} = -\frac{2j+3}{2j+1} \, a_{2j-1}$$

Multiplying together both sides of the equations for even values of k and dividing through by $a_2 \cdot a_4 \cdot a_6 \ldots a_{2j-2}$ gives

$$a_{2j} = \frac{[4 \cdot 6 \cdot 8 \ldots 2(j+1)]}{2 \cdot 4 \cdot 6 \ldots 2j} \, a_0$$

which becomes, upon removing the common factors of the numerator and denominator,

$$a_{2j} = (j+1)a_0 \qquad \text{for } j=1, 2, \ldots \qquad (6\text{--}75)$$

And upon proceeding the same way for the odd values of k, we obtain

$$a_{2j+1} = \frac{2j+3}{3} a_1 \qquad \text{for } j=1, 2, \ldots \qquad (6\text{--}76)$$

Since equations (6–75) and (6–76) show that the coefficients of even and odd powers of z are given by different expressions, it is convenient to first rearrange the series (6–73) into two series, one containing the even powers of z and the others containing the odd powers. This is legitimate since every rearrangement of a convergent powers series converges to the same sum. Thus, equation (6–73) becomes

$$w = \sum_{j=0}^{\infty} a_{2j} z^{2j} + \sum_{j=0}^{\infty} a_{2j+1} z^{2j+1}$$

And upon substituting equations (6–75) and (6–76) into this expression, we obtain the general solution

$$w = a_0 \sum_{j=0}^{\infty} (j+1) z^{2j} + a_1 \sum_{j=1}^{\infty} \left(\frac{2j+3}{3} \right) z^{2j+1}$$

to equation (6–72). It is easy to see that both series converge in the circle $|z| < 1$ and diverge outside this circle.

In both the examples given thus far in this section, explicit expressions for the general terms in the series were obtained. Also the recurrence relations in both examples (see eqs. (6–59) and (6–74)) involved only two different coefficients; whereas, in general, the recurrence relation will involve n different coefficients (see eq. (6–37)). It is, in fact, true in general that there is little hope of obtaining an explicit expression for the solution unless a two-term recurrence relation is obtained. This is not necessarily a limitation on the method because any recurrence relation can be used to successively calculate numerically as many terms of the series as desired. In fact, it is in the cases where many-term recurrence relations occur that the general convergence theorems are particularly useful since, without having an explicit expression for the general term, it is not possible to tell from an infinite series itself whether or not it is convergent.

6.3.2.3 *Case (iii): $\nu = 0, 1, 2, \ldots$ and $a \neq 0$.* — In this case the logarithmic term will be present. Notice that in case (ii) we did not include $\nu = 0$. The reason for this is that when $\nu = 0$ the characteristic exponents are equal and therefore there is only a single recurrence relation. Hence, there can only be one regular solution of the type (6–34) and the logarithmic term must be present in the regular solution (6–70). When $\nu = 1, 2, \ldots$, a will not be equal to zero if, and only if, the recurrence relation (6–71) cannot be solved for a_ν. In this case, no generality will be lost if we set [83] $a = 1$. The solution can always be determined by substituting equation (6–70) into the differential equation, using the fact that w_1 is a solution to simplify the result, and then setting to zero the coefficients of the various powers [84] of $z - z_0$ to calculate the coefficients $a_n^{(2)}$ recursively.

The procedure is best illustrated by considering an example. Thus, the differential equation

$$zw'' - zw' - w = 0 \tag{6–77}$$

has a regular singular point at $z = 0$ and an irregular singular point at $z = \infty$. We know that this equation will possess at least one solution of the form

$$w = \sum_{n=0}^{\infty} a_n z^{n+\rho} \tag{6–78}$$

about the point $z = 0$ and that this solution will converge in the entire complex plane. Upon substituting the expansion (6–78) into equation (6–77), interchanging the order of summation and differentiation, shifting the indices, and collecting terms, we obtain

$$\sum_{n=0}^{\infty} (n+\rho)(n+\rho-1)a_n z^{n+\rho-1} - \sum_{n=1}^{\infty} (n+\rho)a_{n-1} z^{n+\rho-1} = 0$$

Equating the coefficients of like powers of z to zero gives for $n = 0$, since a_0 is arbitrary,

$$\rho(\rho - 1) = 0 \tag{6–79}$$

---

[83] Since w_1 is detemined only to within a constant factor.

[84] The logarithmic terms will always cancel out.

and

$$(n + \rho - 1)a_n = a_{n-1} \qquad \text{for } n = 1, 2, \ldots \tag{6-80}$$

The roots of the indicial equation are $\rho_1 = 1$ and $\rho_2 = 0$; hence, the difference of the roots $\nu = \rho_1 - \rho_2 = 1$ is an integer. First, consider the case where $\rho = \rho_1 = 1$. The recurrence relation is (see eq. (6-76))

$$a_n = \frac{1}{n} a_{n-1} \qquad \text{for } n = 1, 2, \ldots \tag{6-81}$$

After writing out the various terms of equation (6-81), multiplying the corresponding members of these terms together, and dividing out common factors, we get

$$a_n = \frac{1}{1 \cdot 2 \cdot 3 \ldots n} a_0 = \frac{1}{n!} a_0$$

Hence, we find that the solution for $\rho = 1$ is

$$w_1 = a_0 \sum_{n=0}^{\infty} \frac{1}{n!} z^{n+1} \tag{6-82}$$

which can easily be summed to obtain

$$w_1 = a_0 z e^z \tag{6-83}$$

Next, consider the case where $\rho = \rho_2 = 0$. The recurrence relation is $(n - 1)a_n = a_{n-1}$. Since $a_0 \neq 0$, it is clear that this equation cannot be solved for a_1. Hence, the differential equation must possess a logarithmic solution. We, therefore, seek a second solution of the form

$$w_2 = \sum_{n=0}^{\infty} a_n z^n + w_1(z) \ln z \tag{6-84}$$

Upon substituting this expansion into equation (6-77) and recalling that w_1 satisfies the differential equation, we find that the resulting coefficient of $\ln z$

vanishes. Then shifting the index and collecting terms in the summation gives

$$\left(2w_1' - \frac{1}{z}w_1 - w_1 \right) + \sum_{n=1}^{\infty} [n(n-1)a_n - na_{n-1}]z^{n-1} = 0 \qquad (6\text{-}85)$$

Hence, after substituting in equation (6–83), we get

$$a_0(1+z)e^z + \sum_{n=1}^{\infty} [n(n-1)a_n - na_{n-1}]z^{n-1} = 0 \qquad (6\text{-}86)$$

But replacing e^z by its series representation $\sum_{n=0}^{\infty} \frac{1}{n!}z^n$ and then shifting the indices shows that

$$\sum_{n=2}^{\infty} \frac{a_0}{(n-2)!}z^{n-1} + \sum_{n=1}^{\infty} \frac{a_0}{(n-1)!}z^{n-1} + \sum_{n=1}^{\infty} [n(n-1)a_n - na_{n-1}]z^{n-1} = 0$$

Then upon equating to zero the coefficients of like powers of z, we find that

$$a_0 = a_0 \qquad \text{for } n=1$$

and

$$a_n = \frac{a_{n-1}}{n-1} - \frac{a_0}{(n-1)(n-1)!} \qquad \text{for } n=2, 3, 4, \ldots \qquad (6\text{-}87)$$

where a_0 and a_1 are arbitrary. Actually, we could simplify matters by setting $a_1 = 0$, but we shall carry a_1 through as an arbitrary constant in order to demonstrate this.

Upon writing out the first few terms of equation (6–87), we get

$$a_2 = \frac{a_1}{1} - \frac{a_0}{1 \cdot 1!}$$

$$a_3 = \frac{a_2}{2} - \frac{a_0}{2 \cdot 2!} = \frac{a_1}{1 \cdot 2} - \frac{a_0}{2!}\left(1 + \frac{1}{2}\right)$$

$$a_4 = \frac{a_3}{3} - \frac{a_0}{3 \cdot 3!} = \frac{a_1}{1 \cdot 2 \cdot 3} - \frac{a_0}{3!}\left(1 + \frac{1}{2} + \frac{1}{3}\right)$$

Hence, we conclude by induction that

$$a_n = \frac{a_1}{(n-1)!} - \frac{a_0}{(n-1)!} H_{n-1} \qquad \text{for } n = 2, 3, \ldots \qquad (6\text{–}88)$$

where we have defined H_n to be the partial sum of the harmonic series. That is,

$$H_n \equiv \begin{cases} 0 & \text{for } n = 0 \\[2mm] 1 + \dfrac{1}{2} + \dfrac{1}{3} + \ldots + \dfrac{1}{n} = \displaystyle\sum_{k=1}^{n} \frac{1}{k} & \text{for } n = 1, 2, \ldots \end{cases} \qquad (6\text{–}89)$$

Substituting equations (6–83) and (6–88) into equation (6–84) yields

$$w_2 = a_0\left[(ze^z \ln z) + 1 - \sum_{n=2}^{\infty} \frac{1}{(n-1)!} H_{n-1} z^n\right] + a_1 \sum_{n=1}^{\infty} \frac{1}{(n-1)!} z^n$$

which becomes, after shifting the indices in the sums,

$$w_2 = a_0\left[(ze^z \ln z) + 1 - \sum_{n=1}^{\infty} \frac{1}{n!} H_n z^{n+1}\right] + a_1 \sum_{n=0}^{\infty} \frac{1}{n!} z^{n+1}$$

Notice that the second sum is essentially the solution w_1. Hence, the term in square brackets must be a solution which is linearly independent of w_1. This (i.e., the bracketed term) is the solution we would have obtained if we put $a_1 = 0$.

6.3.3 Summary

We have now shown how two linearly independent solutions to the differential equation (6–30) with analytic coefficients can be found in the neighborhood of any regular singular point z_0. First the indicial equation (6–39) is determined. If the difference ν of the two roots of this equation is not an integer (including zero), two solutions of the form (6–34) are obtained, one for each root of the indicial equation. However, if ν is an integer, the recurrence relation (6–37) for the term a_ν must be investigated with $\rho = \rho_2$. It will either automatically be satisfied for any a_ν, or it will be impossible to satisfy for any a_ν. If it is automatically satisfied, a_0 and a_ν will be arbitrary; and we will obtain two linearly independent solutions with $\rho = \rho_2$. If $\nu = 0$ (i.e., $\rho_1 = \rho_2$) or if ν is an integer and it is impossible to satisfy the recurrence relation, the solution will have the form (6–70) and can be found by substituting this form into the differential equation. The point at infinity is treated by making the change in variable $z = 1/\zeta$.

6.3.4 Computation of Indicial Equation

Since the nature of the solutions of equation (6–30) in the neighborhood of a regular singular point z_0 is so dependent on the roots of the indicial equation (characteristic exponents), it is useful to be able to determine this equation without first finding the expansions (6–31) and (6–32) of the coefficients about z_0.

Equation (6–39) shows that this is accomplished once p_0 and q_0 are known. But it follows from equations (6–31) and (6–32) that

$$(z - z_0)p(z) = p_0 + \sum_{k=1}^{\infty} p_k(z - z_0)^k$$

$$(z - z_0)^2 q(z) = q_0 + \sum_{k=1}^{\infty} q_k(z - z_0)^k$$

Hence,

$$p_0 = \lim_{z \to z_0} [(z - z_0)p(z)] \qquad \text{and} \qquad q_0 = \lim_{z \to z_0} [(z - z_0)^2 q(z)]$$

Thus, for example, the equation $z^2 w'' + z(2 + 3z)w' + (1 - z)w = 0$ has a regular singular point at $z = 0$ and

$$p(z) = \frac{2 + 3z}{z} \qquad \text{and} \qquad q(z) = \frac{1 - z}{z^2}$$

Hence,

$$p_0 = \lim_{z \to 0} z \left(\frac{2 - 3z}{z} \right) = 2 \qquad \text{and} \qquad q_0 = \lim_{z \to 0} z^2 \left(\frac{1 - z}{z^2} \right) = 1$$

and the indicial equation (6–39) is

$$\rho^2 + (p_0 - 1)\rho + q_0 = \rho^2 + \rho + 1 = 0$$

If the regular singular point z_0 is the point at infinity, we make the change in variable $z = 1/\zeta$ and investigate the point $\zeta = 0$. This leads to the relations

$$p_0 = \lim_{\zeta \to 0} \left\{ \zeta \left[\frac{2}{\zeta} - \frac{1}{\zeta^2} p \left(\frac{1}{\zeta} \right) \right] \right\}$$

$$q_0 = \lim_{\zeta \to 0} \zeta^2 \left[\frac{1}{\zeta^4} q \left(\frac{1}{\zeta} \right) \right]$$

And upon changing back to the original variable z, they become

$$p_0 = \lim_{z \to \infty} [2 - zp(z)] \qquad \text{and} \qquad q_0 = \lim_{z \to \infty} z^2 q(z)$$

CHAPTER 7

Riemann-Papperitz Equation and the Hypergeometric Equation

7.1 THE FUCHSIAN EQUATION

Having shown how to construct solutions to a differential equation with analytic coefficients about its regular singular points, it is natural to study those equations whose only singularities (in the extended plane) are regular singular points. Such equations are called *Fuchsian equations*. We shall restrict our attention to Fuchsian equations of the second order; that is, equations of the form

$$w'' + p(z)w' + q(z)w = 0 \qquad (7\text{--}1)$$

This equation can have, at most, a finite number of singular points. In order to prove this, notice that since the only singularities of the coefficients p and q in the extended plane are poles, these functions must be rational (see section 5.4). But since the number of singularities of a rational function which occur at finite points of the plane is equal to the number of distinct zeros of its polynomial denominator, it follows that p and q have, at most, a finite number of singular points. However, every second-order equation must have at least one singular point. For if equation (7–1) had no singularities in the extended plane, the coefficients p and q would be everywhere analytic; and, hence, by Liouville's theorem, they would be constants. However, even if the constant value of p were zero, the coefficient $2z - z^2 p(z)$ of the transformed equation (see eq. (6–28)) would still have a simple pole at infinity. But this would contradict the assumption that equation (7–1) had no singularities. Hence, we conclude that this equation must have at least one singularity.

Since the rational function p has, at most, simple poles and the rational function q has, at most, poles of order 2, it is clear that any Fuchsian equation with not more than m singular points in the finite plane must have coefficients of the form

$$p(z) = \frac{P(z)}{(z-z_1)(z-z_2) \ldots (z-z_m)}$$

$$q(z) = \frac{Q(z)}{(z-z_1)^2(z-z_2)^2 \ldots (z-z_m)^2}$$

where P and Q are polynomials.

7.1.1 Fuchsian Equations With, at Most, Two Singular Points

There are two possible forms which a Fuchsian equation with, at most, two singular points can have. First, consider the equation which has, at most, one singular point in the finite plane (with a possible singular point at infinity). Its coefficients must take the form

$$p(z) = \frac{P(z)}{z-a} \qquad q(z) = \frac{Q(z)}{(z-a)^2}$$

But if the coefficients of the transformed equation $2z - z^2 p$ and $z^4 q$ are to have at most a simple pole and, at most, a pole of the order of 2, at infinity the polynomials P and Q must both be constants, say A and B, respectively. Hence, the differential equation must be of the form

$$w'' + \frac{A}{z-a} w' + \frac{B}{(z-a)^2} w = 0 \qquad\qquad (7\text{--}2)$$

Next, consider the case where the singular points are both in the finite plane. Then the coefficients must be of the form

$$p = \frac{P(z)}{(z-a)(z-b)} \qquad q = \frac{Q(z)}{(z-a)^2(z-b)^2}$$

But the coefficients $2z - z^2 p$ and $z^4 q$ of the transformed equation will be analytic at $z = \infty$ only if there are constants C and D such that $P = 2z + C$ and $Q = D$. Hence, in this case the differential equation must be of the form

$$w'' + \frac{2z + C}{(z-a)(z-b)} w' + \frac{D}{(z-a)^2 (z-b)^2} w = 0 \qquad (7\text{--}3)$$

Thus, a Fuchsian equation which has, at most, two singular points must be either of the form (7–2) or of the form (7–3). It is easy to verify that the general solution of equation (7–2) is

$$w = \begin{cases} C_1 (z-a)^{\rho_1} + C_2 (z-a)^{\rho_2} & \text{if } (A-1)^2 - 4B \neq 0 \\ \\ (z-a)^{\rho_1} [C_1 + C_2 \ln(z-a)] & \text{if } (A-1)^2 - 4B = 0 \end{cases}$$

where C_1 and C_2 are arbitrary constants and ρ_1 and ρ_2 are the roots of the indicial equation $\rho^2 + (A-1)\rho + B = 0$. Since the change in independent variable $t = 1/(z-b)$ transforms equation (7–3) into an equation of the form (7–2), it follows that the general solution to equation (7–3) can also be expressed in terms of elementary functions.

Thus, the solutions of any Fuchsian equation which has no more than two singular points in the extended plane are elementary functions.

7.1.2 Fuchsian Equations With, at Most, Three Singular Points

In order to obtain a Fuchsian equation whose solution is not elementary, we must consider an equation which can have three regular singular points. If we require first that these three points, say a, b, and c, all lie in the finite plane, the coefficients of the differential equation must have the form

$$p(z) = \frac{P(z)}{(z-a)(z-b)(z-c)} \qquad q(z) = \frac{Q(z)}{(z-a)^2(z-b)^2(z-c)^2}$$

where P and Q are polynomials. And since the point $z = \infty$ is to be an ordinary point, the coefficients of the transformed equation

$$2z - \frac{z^2 P}{(z-a)(z-b)(z-c)} \qquad \text{and} \qquad \frac{z^4 Q}{(z-a)^2(z-b)^2(z-c)^2}$$

203

must be analytic at $z = \infty$. But this can occur only if P and Q are quadratic in z and the coefficient of z^2 in P is 2. Since the degree of the numerator is less than the degree of the denominator, it follows from the elementary theory of partial fractions that there exist constants $A_a, A_b, A_c, B_a, B_b,$ and B_c such that

$$\frac{P(z)}{(z-a)(z-b)(z-c)} = \frac{A_a}{z-a} + \frac{A_b}{z-b} + \frac{A_c}{z-c}$$

and

$$\frac{Q(z)}{(z-a)(z-b)(z-c)} = \frac{B_a}{z-a} + \frac{B_b}{z-b} + \frac{B_c}{z-c}$$

The differential equation, therefore, takes the form

$$\frac{d^2w}{dz^2} + \left(\frac{A_a}{z-a} + \frac{A_b}{z-b} + \frac{A_c}{z-c}\right)\frac{dw}{dz}$$

$$+ \left(\frac{B_a}{z-a} + \frac{B_b}{z-b} + \frac{B_c}{z-c}\right)\frac{w}{(z-a)(z-b)(z-c)} = 0 \qquad (7\text{-}4)$$

And since the coefficient of z^2 in $P(z)$ must equal 2, it follows that

$$A_a + A_b + A_c = 2 \qquad (7\text{-}5)$$

Otherwise the constants $A_a, A_b, A_c, B_a, B_b,$ and B_c are arbitrary.

Now at each of the points a, b, and c there is a pair of characteristic exponents, say (α', α''), (β', β''), and (γ', γ''),[85] respectively. In order to find a relationship between these exponents and the A's and B's, notice that since

$$p_0 = \lim_{z \to a} (z-a)p(z) = \lim_{z \to a}\left[A_a + A_b\left(\frac{z-a}{z-b}\right) + A_c\left(\frac{z-a}{z-c}\right)\right] = A_a$$

---

[85] The pairs (α', α''), (β', β''), and (γ', γ'') are solutions of the indicial equations at the points a, b, and c, respectively.

and

$$q_0 = \lim_{z \to a} (z-a)^2 q(z) = \lim_{z \to a} \left[B_a + B_b \left(\frac{z-a}{z-b} \right) + B_c \left(\frac{z-a}{z-c} \right) \right] \frac{1}{(z-b)(z-c)}$$

$$= \frac{B_a}{(a-b)(a-c)}$$

at $z = a$, the indicial equation at this point is

$$\rho^2 + (A_a - 1)\rho + \frac{B_a}{(a-b)(a-c)} = 0$$

But it follows from the properties of the roots of quadratic equations that the roots α' and α'' of this equation satisfy the relations

$$\alpha' + \alpha'' = 1 - A_a \qquad \alpha'\alpha'' = \frac{B_a}{(a-b)(a-c)}$$

And we can show in exactly the same way that

$$\beta' + \beta'' = 1 - A_b \qquad \beta'\beta'' = \frac{B_b}{(b-a)(b-c)}$$

$$\gamma' + \gamma'' = 1 - A_c \qquad \gamma'\gamma'' = \frac{B_c}{(c-a)(c-b)}$$

Upon using these relations to eliminate the A's and B's in equation (7–4), we arrive at the *Riemann-Papperitz* equation

$$\frac{d^2w}{dz^2} + \left(\frac{1-\alpha'-\alpha''}{z-a} + \frac{1-\beta'-\beta''}{z-b} + \frac{1-\gamma'-\gamma''}{z-c} \right) \frac{dw}{dz} + \left[\frac{\alpha'\alpha''\,(a-b)\,(a-c)}{z-a} \right.$$

$$\left. + \frac{\beta'\beta''\,(b-a)\,(b-c)}{z-b} + \frac{\gamma'\gamma''\,(c-a)\,(c-b)}{z-c} \right] \frac{w}{(z-a)\,(z-b)\,(z-c)} = 0$$

$$(7-6)$$

Eliminating the A's in equation (7–5) shows that the characteristic exponents at the singular points of equation (7–6) satisfy the relation

$$\alpha' + \alpha'' + \beta' + \beta'' + \gamma' + \gamma'' = 1 \qquad (7\text{–}7)$$

More generally, the sum of all the characteristic exponents at the singular points of any second-order Fuchsian equation is two less than the number of such points. This relation is called the *Fuchsian invariant for the equation of order 2.*

In a similar way, it can be shown that if a Fuchsian equation with, at most, three regular singular points, has one of these points, say c, located at $z = \infty$, this equation must be of the form

$$\frac{d^2w}{dz^2} + \left(\frac{1 - \alpha' - \alpha''}{z - a} + \frac{1 - \beta' - \beta''}{z - b} \right) \frac{dw}{dz}$$

$$+ \left[\frac{\alpha' \alpha'' \, (a - b)}{z - a} + \frac{\beta' \beta'' \, (b - a)}{z - b} + \gamma' \gamma'' \right] \frac{w}{(z - a) \, (z - b)} = 0 \qquad (7\text{–}8)$$

where γ' and γ'' still denote the characteristic exponents at c and equation (7–7) still holds. Since equation (7–8) can be obtained by formally taking the limit $c \to \infty$ in equation (7–6), we can say that equation (7–6) holds even when $c = \infty$. Then with this understanding, equation (7–6) is the most general Fuchsian equation which has, at most, three singular points.

7.2 RIEMANN P-SYMBOL

When w is a solution of the Riemann-Papperitz equation whose singularities occur only at the distinct points $a, b,$ and c, we shall sometimes write

$$w = P \begin{pmatrix} a & b & c & \\ \alpha' & \beta' & \gamma' & z \\ \alpha'' & \beta'' & \gamma'' & \end{pmatrix} \qquad (7\text{–}9)$$

The right side of equation (7–9) is called the *Riemann P-symbol.* The characteristic exponents of the differential equation are listed below the correspond-ing singular point. (Any of the singular points can be at infinity.) These ex-ponents must, of course, satisfy the Fuchsian invariant

$$\alpha' + \alpha'' + \beta' + \beta'' + \gamma' + \gamma'' = 1$$

Finally, the independent variable is located in the fourth column.

If we interchange any of the first three columns in the P-symbol (7–9), the notation still refers to the same equation. There are $3! = 6$ ways in which this can be done. Similarly, interchanging the order of the exponents in any given column leaves the meaning of the symbol unchanged. Thus, for each arrangement of the points a, b, and c, there are eight arrangements of the exponents. Hence, there are $6 \times 8 = 48$ different P-symbols which refer to the same equation and, therefore, have the same meaning.

We shall now show that

$$P\begin{pmatrix} a & b & c & \\ \alpha' & \beta' & \gamma' & z \\ \alpha'' & \beta'' & \gamma'' & \end{pmatrix} = \left(\frac{z-a}{z-c}\right)^k P\begin{pmatrix} a & b & c & \\ \alpha'-k & \beta' & \gamma'+k & z \\ \alpha''-k & \beta'' & \gamma''+k & \end{pmatrix} \quad (7\text{--}10)$$

when a, b, and c are finite and

$$P\begin{pmatrix} a & b & \infty & \\ \alpha' & \beta' & \gamma' & z \\ \alpha'' & \beta'' & \gamma'' & \end{pmatrix} = (z-a)^k P\begin{pmatrix} a & b & \infty & \\ \alpha'-k & \beta' & \gamma'+k & z \\ \alpha''-k & \beta'' & \gamma''+k & \end{pmatrix} \quad (7\text{--}11)$$

when $c = \infty$.

Equation (7–10) means that if w is a solution of a Riemann-Papperitz equation and if we make the change of dependent variable

$$w = \left(\frac{z-a}{z-c}\right)^k u \quad (7\text{--}12)$$

in this equation, then the transformed equation (which has u as the dependent variable) is also a Riemann-Papperitz equation and the location of the singular points is left unchanged by this transformation. However, the characteristic exponents at the points a and c are altered in the manner indicated by the symbols. Equation (7–11) is the corresponding transformation for the case where one of the singular points is at infinity.

In order to prove that equation (7–10) holds, we first substitute equation (7–12) into equation (7–6). Then upon noting that

$$w' = \left(\frac{z-a}{z-c}\right)^k \left[u' + \frac{k(a-c)}{(z-a)(z-c)}u\right]$$

$$w'' = \left(\frac{z-a}{z-c}\right)^k \left[u'' + 2k\left(\frac{1}{z-a} - \frac{1}{z-c}\right)u' + \frac{k(a-c)}{(z-a)(z-c)}\left(\frac{k-1}{z-a} - \frac{k+1}{z-c}\right)u\right]$$

and

$$\frac{1}{(z-a)(z-c)(z-\lambda)} = \frac{1}{(z-a)(z-b)(z-c)}\left(1 + \frac{\lambda-b}{z-\lambda}\right)$$

for $\lambda = a, b$, or c

we find after combining terms that

$$\frac{d^2u}{dz^2} + \left(\frac{1-\alpha'-\alpha''+2k}{z-a} + \frac{1-\beta'-\beta''}{z-b} + \frac{1-\gamma'-\gamma''-2k}{z-c}\right)\frac{du}{dz}$$

$$+ \left[\frac{(\alpha'-k)(\alpha''-k)(a-b)(a-c)}{z-a} + \frac{\beta'\beta''(b-a)(b-c)}{z-b}\right.$$

$$\left. + \frac{(\gamma'+k)(\gamma''+k)(c-a)(c-b)}{z-c}\right]\frac{u}{(z-a)(z-b)(z-c)} = 0$$

But comparing this with equations (7–6) and (7–9) shows that equation (7–10) holds. Equation (7–11) follows from equation (7–8) in the same way.

Having considered the transformations of the Riemann-Papperitz equations under a change of dependent variable, we shall now show how this equation transforms under a change of independent variable. In this case, the change of variable which transforms a Riemann-Papperitz equation into another Riemann-Papperitz equation is the nonsingular linear fractional transformation

$$t = \frac{Az+B}{Cz+D} \qquad \text{where } AD - BC \neq 0 \tag{7-13}$$

studied in chapter 5.

It is again convenient to describe this transformation in terms of the Riemann P-symbol. Thus, if t is the change of variable (7–13), then

$$P\begin{pmatrix} a & b & c & \\ \alpha' & \beta' & \gamma' & z \\ \alpha'' & \beta'' & \gamma'' & \end{pmatrix} = P\begin{pmatrix} a_1 & b_1 & c_1 & \\ \alpha' & \beta' & \gamma' & t \\ \alpha'' & \beta'' & \gamma'' & \end{pmatrix} \tag{7-14}$$

where a_1, b_1, and c_1 are the images of the points a, b, and c, respectively, under equation (7–13). For example,

$$a_1 = \frac{Aa+B}{Ca+D}$$

In order to prove equation (7–14) we first recall that (see section 5.3) performing the transformation (7–13) is equivalent to performing in succession not more than four elementary transformations, each of which has one of the forms

$$t = lz \tag{7-15}$$

$$t = k + z \tag{7-16}$$

$$t = \frac{1}{z} \tag{7-17}$$

It can now be verified, by substituting in turn each of equations (7–15) to (7–17) into equation (7–6), that in each case the transformed equation is a Riemann-Papperitz equation with the same characteristic exponents as equation (7–6) and that the singular points of the transformed equation are the images under the transformation of the original singular points. Hence, equation (7–14) holds for each of the elementary transformations (7–15) to (7–17). But since the general nonsingular linear fractional transformation is equivalent to performing a succession of these transformations, it is clear that equation (7–12) must hold for this transformation also.

7.3 TRANSFORMATION OF RIEMANN-PAPPERITZ EQUATION INTO HYPER-GEOMETRIC EQUATION

We shall now show that the general Riemann-Papperitz equation (7–6) can always be transformed by the application of a number of transformations (each of which has one of the forms (7–10), (7–11), and (7–14)) into a certain standard equation (canonical form). The usefulness of this result is principally due to the fact that the transformed equation, which has a regular singular point at the origin, has a two-term recurrence relation at this point. As we have seen, this allows us to find an explicit expression for the series solution.

In order to accomplish this reduction let w be any solution to equation (7–6). Then

$$
w = P \begin{pmatrix} a & b & c & \\ \alpha' & \beta' & \gamma' & z \\ \alpha'' & \beta'' & \gamma'' & \end{pmatrix}
$$

where we suppose that the notation has been so arranged that a and b are always finite. We first introduce the new dependent variable v by

$$
w = \begin{cases} \left(\dfrac{z-a}{z-c}\right)^{\alpha'} v & \text{if } c \neq \infty \\[2mm] (z-a)^{\alpha'} v & \text{if } c = \infty \end{cases} \tag{7–18}
$$

and then use equation (7–10) or (7–11) to show that

$$v = P \begin{pmatrix} a & b & c & \\ 0 & \beta'' & \gamma' + \alpha' & z \\ \alpha'' - \alpha' & \beta'' & \gamma + \alpha' & \end{pmatrix} \qquad (7\text{-}19)$$

We again change the dependent variable by using the transformation

$$v = \begin{cases} \left(\dfrac{z-b}{z-c}\right)^{\beta'} u & \text{if } c \neq \infty \\[3mm] (z-b)^{\beta'} u & \text{if } c = \infty \end{cases} \qquad (7\text{-}20)$$

and then apply equation (7–10) or (7–11) to equation (7–19) to show that

$$u = P \begin{pmatrix} a & b & c & \\ 0 & 0 & \gamma' + \alpha' + \beta' & z \\ \alpha'' - \alpha' & \beta'' - \beta' & \gamma'' + \alpha' + \beta' & \end{pmatrix} \qquad (7\text{-}21)$$

Now let us use the linear fractional transformation (discussed in section 5.3) which maps the points a, b, and c into the points $0, 1$, and ∞, respectively, to change the independent variable. Thus, a new independent variable t is introduced by

$$t = \left(\frac{z-a}{b-a}\right)\left(\frac{b-c}{z-c}\right) \qquad (7\text{-}22)$$

(When $c = \infty$, we understand t to be the limit of this expression as $c \to \infty$, which is simply the first term in the right-hand member.) And in view of equation (7–14) it follows from equations (7–21) and (7–22) that

$$u = P \begin{pmatrix} 0 & 1 & \infty & \\ 0 & 0 & \gamma' + \alpha' + \beta' & t \\ \alpha'' - \alpha' & \beta'' - \beta' & \gamma'' + \alpha' + \beta' & \end{pmatrix} \qquad (7\text{-}23)$$

The transformations (7–18). (7–20), and (7–22) can now be collected into the single formula

$$w(z) = \begin{cases} \left(\dfrac{z-a}{z-c}\right)^{\alpha'} \left(\dfrac{z-b}{z-c}\right)^{\beta'} u(t) & t = \left(\dfrac{z-a}{b-a}\right)\left(\dfrac{b-c}{z-c}\right) & \text{for } c \neq \infty \\[2em] (z-a)^{\alpha'} (z-b)^{\beta'} u(t) & t = \dfrac{z-a}{b-a} & \text{for } c = \infty \end{cases}$$

(7–24)

The function u given in equation (7–23) is the solution to a Riemann-Papperitz equation which has only four nonzero exponents. Since the sum of these exponents is unity, they can be characterized by three constants, say α, β, and γ. It is customary to define these constants by

$$\alpha = \alpha' + \beta' + \gamma' \qquad \beta = \alpha' + \beta' + \gamma'' \qquad \gamma = 1 + \alpha' - \alpha''$$

(7–25)

Then equation (7–23) can be written as

$$u = P \begin{pmatrix} 0 & 1 & \infty & \\ 0 & 0 & \alpha & t \\ 1-\gamma & \gamma-\alpha-\beta & \beta & \end{pmatrix}$$

(7–26)

But comparing equations (7–8), (7–9), and (7–26) shows that the Fuchsian equation satisfied by (7–26) is

$$\frac{d^2u}{dt^2} + \left[\frac{1-(1-\gamma)}{t} + \frac{1-(\gamma-\alpha-\beta)}{t-1}\right]\frac{du}{dt} + \left[\frac{0}{t} + \frac{0}{t-1} + \alpha\beta\right]\frac{u}{t(t-1)} = 0$$

And upon rearranging this equation slightly, we obtain the *hypergeometric equation of Gauss*,

$$t(1-t)\frac{d^2u}{dt^2} + [\gamma - (\alpha+\beta+1)\,t]\frac{du}{dt} - \alpha\beta u = 0$$

(7–27)

Its solutions, which will be obtained subsequently, have been extensively studied.

We have, therefore, shown that *any Riemann-Papperitz equation can be transformed into the hypergeometric equation (7–27) by introducing a change of variable of the type (7–24).*

Hence, once the solutions to the hypergeometric equation (7–27) have been found, we can find the solutions w to any Riemann-Papperitz equation simply by using the transformation (7–24) to express them in terms of the solutions to the hypergeometric equation.

However, in any specific case, it is usually more convenient to rederive this transformation by using the Riemann P-symbol. Thus, consider the Riemann-Papperitz equation

$$2z^2(z-2)w'' - z(3z-2)w' + 2(z-1)w = 0 \qquad (7\text{–}28)$$

For this equation,

$$p(z) = -\frac{3z-2}{2z(z-2)} \qquad\qquad q(z) = \frac{z-1}{z^2(z-2)}$$

Hence, its singularities are located at the points $z=0$, $z=2$, and $z=\infty$.

At $z=0$

$$p_0 = \lim_{z\to 0}\left[-\frac{3z-2}{2(z-2)}\right] = -\frac{1}{2} \qquad\qquad q_0 = \lim_{z\to 0}\frac{z-1}{z-2} = \frac{1}{2}$$

and, therefore, the indicial equation at this point is (see eq. (6–39))

$$\rho^2 - \frac{3}{2}\rho + \frac{1}{2} = 0$$

The roots of this equation are 1 and 1/2.

At $z=2$

$$p_0 = \lim_{z\to 2}\left(-\frac{3z-2}{2z}\right) = -1 \qquad\qquad q_0 = \lim_{z\to 2}\frac{(z-2)(z-1)}{z^2} = 0$$

and, therefore, the indicial equation is

$$\rho^2 - 2\rho = 0$$

The roots of this equation are 2 and 0.

At $z = \infty$

$$p_0 = \lim_{z \to \infty} [2 - zp(z)] = \lim_{z \to \infty} \left[2 + \frac{3z-2}{2(z-2)} \right] = \frac{7}{2}$$

$$q_0 = \lim_{z \to \infty} z^2 q(z) = \lim_{z \to \infty} \frac{z-1}{z-2} = 1$$

and, therefore, the indicial equation is

$$\rho^2 + \frac{5}{2}\rho + 1 = 0$$

The roots of this equation are $-1/2$ and -2. Hence, the solutions to equation (7–38) can be denoted by

$$w = P \begin{pmatrix} 0 & 2 & \infty & \\ 1 & 2 & -1/2 & z \\ 1/2 & 0 & -2 & \end{pmatrix}$$

It now follows from equation (7–11) that

$$w = z^{1/2} P \begin{pmatrix} 0 & 2 & \infty & \\ 0 & 0 & 0 & z \\ 1/2 & 2 & -3/2 & \end{pmatrix}$$

And it can be concluded from equation (7–14) that

$$w = z^{1/2} P \begin{pmatrix} 0 & 1 & \infty & \\ 0 & 0 & 0 & t \\ 1/2 & 2 & -3/2 & \end{pmatrix}$$

where $t = z/2$. But upon comparing this with equation (7–26), we see that

$$\alpha = 0 \qquad \beta = -\frac{3}{2} \qquad \gamma = \frac{1}{2} \qquad (7\text{–}29)$$

Thus, the solutions to equation (7–28) can be expressed in terms of the solutions to the hypergeometric equation (7–27) with the constants α, β, and γ given by equation (7–29).

7.4 HYPERGEOMETRIC FUNCTIONS

It remains to determine the solutions to the hypergeometric equation (7–27). We shall first seek a solution about the regular singular point $z = 0$. Since the P-symbol for equation (7–27) is given by equation (7–26), it follows that the characteristic exponents at $z = 0$ are 0 and $1 - \gamma$. Suppose first that γ is neither zero nor a negative integer. Then the methods developed in chapter 6 can be used to obtain a power series solution [86] of the form

$$u = \sum_{n=0}^{\infty} a_n t^n \qquad (7\text{–}30)$$

which corresponds to the characteristic exponent zero. Upon substituting this into equation (7–27), collecting terms with like powers of t, and shifting the index n, the lowest one present, we obtain

$$\sum_{n=0}^{\infty} n(\gamma + n - 1) a_n t^{n-1} - \sum_{n=1}^{\infty} (\alpha + n - 1)(\beta + n - 1) a_{n-1} t^{n-1} = 0$$

Hence, the $n = 0$ term shows that a_0 is arbitrary and we obtain the two-term recurrence relation

$$a_n = \frac{(\alpha + n - 1)(\beta + n - 1)}{n(\gamma + n - 1)} a_{n-1} \qquad \text{for } n = 1, 2, \ldots \qquad (7\text{–}31)$$

where the division by $n(\gamma + n - 1)$ is permitted since we have required that γ be neither zero nor a negative integer and, therefore, the denominator can never vanish. Upon writing out equation (7–31) for successive values of n, multiplying the corresponding members of these equations together, and dividing

---

[86] Recall that it was shown in section 6.3 that the power series solution corresponding to the largest characteristic exponent always exists.

out common factors, we find that

$$a_n = \frac{[\alpha(\alpha+1) \ldots (\alpha+n-1)][\beta(\beta+1) \ldots (\beta+n-1)]}{n![\gamma(\gamma+1) \ldots (\gamma+n-1)]} a_0$$

And by using the generalized factorial function defined in equation (5–33), this can be written in the more concise form

$$a_n = \frac{(\alpha)_n (\beta)_n}{n!(\gamma)_n} a_0$$

Hence, the solution (7–30) is

$$u = a_0 F(\alpha, \beta; \gamma; t) \tag{7–32}$$

where we have put

$$F(\alpha, \beta; \gamma; t) \equiv \sum_{n=0}^{\infty} \frac{(\alpha)_n (\beta)_n}{n!(\gamma)_n} t^n \qquad \text{for } \gamma \neq 0, -1, -2, \ldots \tag{7–33}$$

The function $F(\alpha, \beta; \gamma; t)$ is called the *hypergeometric function*, and the notation used herein is universally accepted. The reason for calling this function the hypergeometric function is that when $\alpha=1$ and $\beta=\gamma$, this series (7–33) reduces to the geometric series. Thus,

$$F(1, \gamma; \gamma; t) = \frac{1}{1-t}$$

In fact, it follows from the binomial theorem that

$$F(\alpha, \gamma; \gamma; t) = F(\gamma, \alpha; \gamma; t) = \frac{1}{(1-t)^\alpha} \tag{7–34}$$

for any number α.

Notice that if either α or β is zero or a negative integer, the series (7–33) terminates after a finite number of terms and, therefore, represents an entire function. In all other cases, a simple application of the ratio test (ref. 23) will suffice to show that it converges in the circle $|t| < 1$ and diverges outside this circle.

We have, therefore, shown that, provided γ is neither zero nor a negative integer,

$$u = F(\alpha, \beta; \gamma; t) \tag{7–35}$$

is a solution to the hypergeometric equation

$$t(1-t)u'' + [\gamma - (\alpha + \beta + 1)t]u' - \alpha\beta u = 0 \tag{7–36}$$

at least in the circle $|t| < 1$.

Since equation (7–36) is a Riemann-Papperitz equation, its solutions u can be denoted by the P-symbol

$$u = P \begin{pmatrix} 0 & 1 & \infty & \\ 0 & 0 & \alpha & t \\ 1-\gamma & \gamma-\alpha-\beta & \beta & \end{pmatrix} \tag{7–37}$$

Also, inasmuch as the order in which the exponents are listed in a given column is immaterial, we can interchange those in the first column to obtain

$$u = P \begin{pmatrix} 0 & 1 & \infty & \\ 1-\gamma & 0 & \alpha & t \\ 0 & \gamma-\alpha-\beta & \beta & \end{pmatrix}$$

Hence, it follows from equation (7–11) that

$$u = t^{1-\gamma} P \begin{pmatrix} 0 & 1 & \infty & \\ 0 & 0 & \alpha+1-\gamma & t \\ \gamma-1 & \gamma-\alpha-\beta & \beta+1-\gamma & \end{pmatrix} \qquad (7\text{--}38)$$

Notice that the P-symbol in equation (7–38) is in canonical form. It, therefore, denotes a solution to the hypergeometric equation whose exponents at the points 0, 1, and ∞ are $0, \gamma-1$: $0, \gamma-\alpha-\beta$; and $\alpha+1-\gamma, \beta+1-\gamma$, respectively. And we can choose this solution to be the hypergeometric function F. Hence, upon comparing equations (7–35) and (7–37) with the P-symbol in equation (7–38), we find that

$$u = t^{1-\gamma} F(\alpha+1-\gamma, \beta+1-\gamma; 2-\gamma; t) \qquad (7\text{--}39)$$

is also a solution to equation (7–36) provided, of course, $2-\gamma$ is neither zero nor a negative integer. Since the hypergeometric series converges for $|t| < 1$, the function (7–39) must satisfy the differential equation in this region. Notice that if $\gamma=1$, the solutions (7–35) and (7–39) are identical. In fact, equation (7–37) shows that, in this case, equation (7–36) has equal exponents at $t=0$ and, therefore, one of the solutions composing the canonical basis at $t=0$ must contain logarithmic terms. Next, if γ is any integer other than unity, either γ or $2-\gamma$ is either zero or a negative integer. Thus, either the hypergeometric function (7–35) or the hypergeometric function (7–39) is undefined. The remaining equation provides a solution to the differential equation. Finally, if γ is not an integer, both equations (7–35) and (7–39) are solutions to the differential equation in the circle $|t| < 1$. Since the series expansion for equation (7–35) begins with t^0 and that for equation (7–39) with $t^{1-\gamma}$, they must also be linearly independent solutions.

Interchanging the exponents in the second column of equation (7–37) leads in the manner described above to the solution

$$u = (t-1)^{\gamma-\alpha-\beta} F(\gamma-\beta, \gamma-\alpha; \gamma; t) \qquad (7\text{--}40)$$

in the domain $|t| < 1$. And interchanging both the exponents in the first column and those in the second column leads to a fourth solution

$$u = t^{1-\gamma}(t-1)^{\gamma-\alpha-\beta} F(1-\beta, 1-\alpha; 2-\gamma; t) \qquad (7\text{--}41)$$

in the domain $|t| < 1$. But, at most, two of these four solutions can be linearly independent. If γ is not an integer, there exists a linear relation connecting any three of them. And, if γ is an integer, the two solutions which exist can differ only by a multiplicative constant.

Another set of four solutions can be obtained by interchanging the points 0 and 1 by means of the transformation $z = 1 - t$. Thus, for example, equation (7–37) is equivalent to the equation

$$u = P \begin{pmatrix} 1 & 0 & \infty & \\ \gamma - \alpha - \beta & 1 - \gamma & \alpha & t \\ 0 & 0 & \beta & \end{pmatrix}$$

Upon applying equation (7–11) to this equation twice in succession, we obtain

$$u = (t-1)^{\gamma-\alpha-\beta} t^{1-\gamma} P \begin{pmatrix} 1 & 0 & \infty & \\ 0 & 0 & 1-\beta & t \\ \alpha+\beta-\gamma & \gamma-1 & 1-\alpha & \end{pmatrix}$$

It now follows from equations (7–13) and (7–14) that the transformation $z = 1 - t$ maps the points 1, 0, and ∞ into the points 0, 1, and ∞, respectively, and leads to the expression

$$u = (t-1)^{\gamma-\alpha-\beta} t^{1-\gamma} P \begin{pmatrix} 0 & 1 & \infty & \\ 0 & 0 & 1-\beta & 1-t \\ \alpha+\beta-\gamma & \gamma-1 & 1-\alpha & \end{pmatrix}$$

Since this P-symbol is in canonical form, it represents a solution to a hypergeometric equation which we can again choose to be F. Thus, we find after comparing exponents with equations (7–35) and (7–37) that

$$u = (t-1)^{\gamma-\alpha-\beta} t^{1-\gamma} F(1-\beta, 1-\alpha; 1+\gamma-\alpha-\beta; 1-t) \qquad (7\text{–}42)$$

Notice that the hypergeometric series in the solution (7–35) converges in the circle $|t| < 1$, while that in the solution (7–42) converges in the circle $|t-1| < 1$. By systematically exploiting the technique used above, Kummer (1810 to 1893) obtained 24 solutions to the hypergeometric equation (7–36), each of which is expressed in terms of the hypergeometric function F. In each of these solutions the variable is one of the quantities t, $1-t$, $(t-1)/t$, $t/(1-t)$, $1/t$, and $(1-t)^{-1}$. Since the function F with variable t converges in the unit circle $|t| < 1$, these solutions will converge in one of the regions shown in figure 7–1. Each series will converge in the appropriate shaded region and diverge outside it unless the series terminates; in which case, it will converge in the entire plane. Of the 24 solutions, four converge in each of the regions shown. Of course, only two solutions can be linearly independent in any given region. Some of the regions shown in the figure overlap. Hence, there are numerous relations among the various 24 solutions obtained by Kummer. These relations not only provide useful identities among hypergeometric functions but also provide formulas which can be used to analytically continue the solutions of the Riemann-Papperitz equation from one region of the plane to another. Thus, it can be shown, for example, that when γ is neither zero nor a negative integer, the identity

$$F(\alpha, \beta; \gamma; t) = (1-t)^{\gamma-\alpha-\beta} F(\gamma-\alpha, \gamma-\beta; \gamma; t) \tag{7–43}$$

holds in the domain $|t| < 1$. An example of an identity which holds in the intersection of the regions shown in figures 7–1(a) and (b) is

$$F(\alpha, \beta; \gamma; t) = (1-t)^{-\alpha} F\left(\alpha, \gamma-\beta; \gamma; \frac{1}{t-1}\right) \qquad \text{for } |t| < 1 \text{ and } |t| < |1-t|$$

It is frequently convenient to express the hypergeometric function as an integral. This can be done when $\mathscr{Re}\,\gamma > \mathscr{Re}\,\beta > 0$ and $|t| < 1$. In order to arrive at this result, we substitute equation (5–35) into equation (7–33) to obtain

$$F(\alpha, \beta; \gamma; t) = \frac{\Gamma(\gamma)}{\Gamma(\beta)} \sum_{n=0}^{\infty} \frac{(\alpha)_n}{n!} \frac{\Gamma(\beta+n)}{\Gamma(\gamma+n)} t^n \tag{7–44}$$

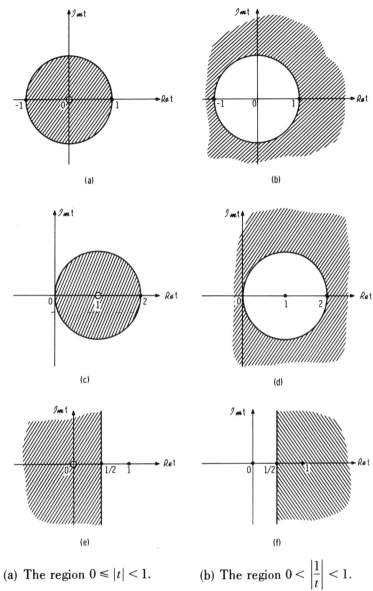

(a) The region $0 \leqslant |t| < 1$. (b) The region $0 < \left|\dfrac{1}{t}\right| < 1$.

(c) The region $0 \leqslant |1-t| < 1$. (d) The region $0 < \left|\dfrac{1}{1-t}\right| < 1$.

(e) The region $0 \leqslant \left|\dfrac{t}{t-1}\right| < 1$. (f) The region $0 \leqslant \left|\dfrac{t-1}{t}\right| < 1$.

FIGURE 7–1.—Regions of convergence for hypergeometric series.

Since it follows from equations (5–37) and (5–38) that

$$\frac{\Gamma\ (\beta+n)\ \Gamma\ (\gamma-\beta)}{\Gamma\ (\gamma+n)}=B\ (\beta+n,\gamma-\beta)=\int_0^1 \tau^{\beta+n-1}\ (1-\tau)^{\gamma-\beta-1}d\tau$$

$$\text{for } \mathscr{R}_e\,\gamma > \mathscr{R}_e\,\beta > 0$$

equation (7–44) can be written as

$$F\ (\alpha,\beta;\gamma;t)=\frac{\Gamma(\gamma)}{\Gamma(\beta)\,\Gamma\ (\gamma-\beta)}\sum_{n=0}^{\infty}\frac{(\alpha)_n}{n!}\,t^n\int_0^1 \tau^{\beta+n-1}(1-\tau)^{\gamma-\beta-1}d\tau$$

Upon interchanging the order of integration and summation (see section 5.6), we find that

$$F\ (\alpha,\beta;\gamma;t)=\frac{\Gamma(\gamma)}{\Gamma(\beta)\,\Gamma\ (\gamma-\beta)}\int_0^1 \tau^{\beta-1}(1-\tau)^{\gamma-\beta-1}\sum_{n=0}^{\infty}\frac{(\alpha)_n}{n!}\,(t\tau)^n\,d\tau$$

But the series can be summed by using equations (7–33) and (7–34) to obtain the integral representation

$$F\ (\alpha,\beta;\gamma;t)=\frac{\Gamma(\gamma)}{\Gamma(\beta)\,\Gamma\ (\gamma-\beta)}\int_0^1 \tau^{\beta-1}(1-\tau)^{\gamma-\beta-1}(1-\tau t)^{-\alpha}d\tau$$

$$\text{for } \mathscr{R}_e\,\gamma > \mathscr{R}_e\,\beta > 0 \text{ and } |t| < 1$$

which was discovered by Euler. Since this integral exists and represents an analytic function for all values of t except where t is real and larger than 1, it provides an analytic continuation of $F(\alpha,\ \beta;\ \gamma;\ t)$ from the unit circle $|t| < 1$ to the entire t-plane cut along the positive real axis. Barnes obtained the more general integral representation

$$F\ (\alpha,\beta;\gamma;t)=\frac{1}{2\pi i}\frac{\Gamma(\gamma)}{\Gamma(\alpha)\,\Gamma(\beta)}\int_{-i\infty}^{i\infty}\frac{\Gamma\ (\alpha+z)\ \Gamma\ (\beta+z)\ \Gamma\ (-z)}{\Gamma\ (\gamma+z\ )}(-t)^z dz$$

where $|\arg\ (-t)| < \pi$ and the integration is to be carried out along any path which lies to the right of the poles of $\Gamma(\alpha+z)\ \Gamma(\beta+z)$ and to the left of the poles of $\Gamma(-z)$.

The six functions $F(\alpha \pm 1, \beta; \gamma; t)$, $F(\alpha, \beta \pm 1, \gamma; t)$, and $F(\alpha, \beta; \gamma \pm 1; t)$ are said to be contiguous to $F(\alpha, \beta; \gamma; t)$. Gauss proved that there is a linear relation between $F(\alpha, \beta; \gamma; t)$ and any two of its contiguous functions, the coefficients being linear polynomials in t. Some examples of these relations are

$$\gamma(1-t)F(\alpha, \beta; \gamma; t) = (\beta - \gamma)tF(\alpha, \beta; \gamma + 1; t) + \gamma F(\alpha - 1, \beta; \gamma; t)$$

$$(\beta - a)F(\alpha, \beta; \gamma; t) = \beta F(\alpha, \beta + 1; \gamma; t) - \alpha F(\alpha + 1, \beta; \gamma; t)$$

The derivative of a hypergeometric function is also a hypergeometric function. In order to show this, notice that the series (7–33) can be differentiated term by term within its circle of convergence to obtain

$$\frac{dF}{dt} = \sum_{n=1}^{\infty} \frac{(\alpha)_n (\beta)_n t^{n-1}}{(\gamma)_n (n-1)!}$$

Upon shifting the index by putting $k = n - 1$, this becomes

$$\frac{dF}{dt} = \sum_{k=0}^{\infty} \frac{(\alpha)_{k+1} (\beta)_{k+1}}{(\gamma)_{k+1} k!} t^k \tag{7–45}$$

But since it follows from definition (5–33) that $(\alpha)_{k+1} = \alpha(\alpha + 1)_k$, $(\beta)_{k+1} = \beta(\beta + 1)_k$, and $(\gamma)_{k+1} = \gamma(\gamma + 1)_k$, equation (7–45) shows that

$$\frac{dF(\alpha, \beta; \gamma; t)}{dt} = \frac{\alpha\beta}{\gamma} F(\alpha + 1, \beta + 1; \gamma + 1; t) \tag{7–46}$$

More generally, the *Jacobi identity*

$$\frac{d^n}{dt^n} [t^{\alpha + n - 1} F(\alpha, \beta; \gamma; t)] = (\alpha)_n t^{\alpha - 1} F(\alpha + n, \beta; \gamma; t) \tag{7–47}$$

can be established by multiplying both sides of the identity by $t^{1-\alpha}$ and then verifying that the left side satisfies the hypergeometric equation with parameters $\alpha + n$, β, γ.

There are many other relations connecting the hypergeometric functions. The reader can find these tabulated in reference 26 and discussed in some detail in reference 27. Many of the known functions, such as the elliptic integrals of the first and second kind, the inverse sine, etc., can be expressed as hypergeometric functions.

7.5 GENERALIZED HYPERGEOMETRIC FUNCTIONS

The concept of hypergeometric function can, itself, be generalized. To this end, notice that in the definition of the hypergeometric function

$$F(\alpha, \beta; \gamma; z) = \sum_{n=0}^{\infty} \frac{(\alpha)_n (\beta)_n}{(\gamma)_n n!} z^n$$

the numerator and denominator of the coefficient of $z^n/n!$ contain only products of generalized factorial functions. Hence, it is natural to define the *generalized hypergeometric function* $_pF_q$ by

$$_pF_q(\alpha_1, \alpha_2, \ldots, \alpha_p; \gamma_1, \gamma_2, \ldots, \gamma_q; z) = \sum_{n=0}^{\infty} \frac{(\alpha_1)_n (\alpha_2)_n \ldots (\alpha_p)_n z^n}{n! (\gamma_1)_n (\gamma_2)_n \ldots (\gamma_q)_n}$$

$$(7-48)$$

provided the series converges.[87] The numbers p and q are nonnegative integers which denote the number of parameters in the numerator and in the denominator, respectively. Thus, the ordinary hypergeometric function F is a $_2F_1$. Since the parameter β cancels out in the numerator and denominator in equation (7-34), we see that

$$\frac{1}{(1-t)^\alpha} = _1F_0(\alpha; -; t)$$

where we indicate that there is no parameter in the denominator by inserting a dash. The function $_1F_1$ is treated in the next chapter. The generalized hypergeometric functions are studied in considerable detail in reference 27.

---

[87] This series will always converge in the unit circle if $q = p + 1$, and it will converge in the entire plane if $q > p + 1$. The other cases are more complex.

7.6 LOGARITHMIC SOLUTION TO HYPERGEOMETRIC EQUATION

We have found two linearly independent solutions to the hypergeometric equation (7–36) in a neighborhood of $t=0$ for the case where γ is not an integer. However, when γ is an integer, only one solution has been found. The remaining solution can be found by the methods of chapter 6. Thus, for example, when γ is a positive integer, $u_1 = F(\alpha, \beta; \gamma; t)$ provides one solution to the hypergeometric equation (7–36); and the methods of chapter 6 show that if neither α nor β is a positive integer smaller than γ, a second linearly independent solution is given by

$$u_2 = F(\alpha, \beta; \gamma; 1) \ln t - \sum_{n=0}^{\overline{\gamma-2}} \frac{n!\,(1-\gamma)_{n+1}}{(1-\alpha)_{n+1}(1-\beta)_{n+1}} t^{-(n+1)}$$

$$+ \sum_{n=1}^{\infty} \frac{(\alpha)_n(\beta)_n}{n!\,(\gamma)_n} \left[H(\alpha, n) + H(\beta, n) - H(\gamma, n) - H(1, n) \right] t^n$$

where the finite sum is to be omitted when $\gamma = 1$ and we have put[88]

$$H(\alpha, n) \equiv \sum_{k=0}^{n-1} \frac{1}{\alpha + k}$$

We can obtain the second solution for the case where γ is zero or a negative integer in the same way by starting with the solution

$$u_1 = t^{1-\gamma} F(\alpha + 1 - \gamma, \beta + 1 - \gamma; 2 - \gamma; t)$$

7.7 SPECIAL RIEMANN-PAPPERITZ EQUATIONS

The classical *Jacobi differential equation*

$$(1-z^2)w'' + [b - a - (a+b+2)z]w' + n(n+a+b+1)w = 0 \quad (7\text{–}49)$$

---

[88] The function defined in eq. (6–89) is identical to $H(1, n)$.

is a particular Riemann-Papperitz equation with singular points at $z = \pm 1, \infty$. It is readily established that in this case

$$
w = P \begin{pmatrix}
1 & -1 & \infty & \\
0 & 0 & -n & z \\
-a & -b & n+a+b+1 &
\end{pmatrix}
\tag{7-50}
$$

This equation can be put into canonical form by the change of independent variable $t = (1-z)/2$. Thus,

$$
w = P \begin{pmatrix}
0 & 1 & \infty & \\
0 & 0 & -n & t' \\
-a & -b & n+a+b+1 &
\end{pmatrix}
$$

Therefore, when a is not a negative integer, a particular solution w_1 to the Jacobi differential equation is

$$
w_1 = F\left(-n,\, n+a+b+1;\, a+1;\, \frac{1-z}{2}\right)
\tag{7-51}
$$

Equation (7–51) satisfies the differential equation within the circle $|1-z| < 2$ for all values of the parameter n. But if n is a nonnegative integer, the hypergeometric series terminates after a finite number of terms. For such values of n, the *Jacobi polynomial* is defined as

$$
P_n^{(a,\,b)}(z) = \frac{(a+1)_n}{n!}\, F\left(-n,\, n+a+b+1;\, a+1;\, \frac{1-z}{2}\right) \quad \text{for } n = 0, 1, 2, \ldots
\tag{7-52}
$$

A convenient expression for these polynomials is provided by the *Rodrigues formula*

$$
P_n^{(a,\,b)}(z) = \frac{(-1)^n}{n!\, 2^n}\, (1-z)^{-a}(1+z)^{-b}\, \frac{d^n}{dz^n}\left[(1-z)^{a+n}(1+z)^{b+n}\right]
\tag{7-53}
$$

This formula was obtained for the special case of Legendre polynomials (introduced below) by Olinde Rodrigues in 1816. In order to derive this result, we use equation (7–34) together with the Jacobi identity (7–47) to show that

$$t^{-a}(1-t)^{-b}\frac{d^n}{dt^n}\left[(1-t)^{b+n}t^{a+n}\right]$$

$$=t^{-a}(1-t)^{-b}\frac{d^n}{dt^n}\left[t^{a+n}F\left(a+1,-n-b;a+1;t\right)\right]$$

$$=(a+1)_n(1-t)^{-b}F\left(a+1+n,-n-b;a+1;t\right) \qquad (7-54)$$

But it follows from equation (7–43) that

$$(1-t)^{-b}F\left(a+1+n,-n-b;a+1;t\right)=F\left(-n,a+1+n+b;a+1;t\right)$$

Hence, after substituting this into equation (7–54) and making the change of variable $t=(1-z)/2$ in the resulting expression, we find, upon comparing the result with definition (7–52), that the Rodrigues formula (7–53) holds.

A number of important equations which have been extensively studied are special cases of Jacobi's equation. Thus, when $a=b$, equation (7–49) reduces to the *ultraspherical differential equation*

$$(1-z^2)\,w''-2\,(a+1)\,zw'+n\,(n+2a+1)\,w=0 \qquad (7-55)$$

A solution to this equation is given by equation (7–51) with $a=b$. But, when n is a nonnegative integer and the series terminates,[89] it is customary to use a different normalization factor from the one used in equation (7–52) and to express the solution in terms of the *Gegenbauer polynomial*

$$C_n^\alpha(z) = \frac{(2\alpha)_n}{\left(\alpha+\dfrac{1}{2}\right)_n}\,P_n^{[\alpha-(1/2),\,\alpha-(1/2)]}(z) \qquad (7-56)$$

---

[89] The polynomial obtained directly from eq. (7–52) with $a=b$ is sometimes called the *ultraspherical polynomial*.

where we have put $\alpha = a + (1/2)$.

The *Tschebychev equation*

$$(1 - z^2)w'' - zw' + n^2 w = 0 \tag{7-57}$$

is, in turn, a special case of the ultraspherical equation (7–55) with $a = -1/2$. One solution to this equation is therefore given by equation (7–51) with $a = b = -1/2$. However, when n is a nonnegative integer, it is again customary to change the normalization and define the *Tschebychev polynomial* (of the first kind) T_n in terms of the Jacobi polynomial (or ultraspherical polynomial) by [90]

$$T_n(z) = \frac{n!}{\left(\dfrac{1}{2}\right)_n} P_n^{(-1/2,\,-1/2)}(z)$$

Another important special case of the ultraspherical equation occurs when we put $a = 0$ to obtain *Legendre's equation*

$$(1 - z^2)\,w'' - 2zw' + n\,(n+1)\,w = 0 \tag{7-58}$$

One solution to this equation is given by equation (7–51) with $a = b = 0$. Thus, we define the *Legendre function of the first kind* P_n by

$$P_n(z) = F\left(-n, n+1; 1; \frac{1-z}{2}\right) \tag{7-59}$$

And since this function is a polynomial when n is a nonnegative integer, it is then called the *Legendre polynomial*. Thus, the Legendre polynomial is a Jacobi polynomial with $a = b = 0$.

A second (suitably normalized) linearly independent solution to equation (7–58) about $z = 1$ is called the *Legendre function of the second kind* and is denoted by Q_n. If n is an integer, Q_n will involve logarithmic terms.

Notice that Legendre's differential equation (7–58) (and, in fact, more generally, the ultraspherical differential equation (7–55)) is invariant when z is replaced by $-z$. This suggests that we transform equation (7–58) by introducing

---

[90] It is not hard to show that T_n satisfies the relation $T_n (\cos \theta) = \cos n\,\theta$, and this relation is often used to define this polynomial.

the new independent variable $t = z^2$. The equation then becomes

$$2t(1-t)\frac{d^2w}{dt^2} + (1-3t)\frac{dw}{dt} + \frac{1}{2}n(n+1)w = 0$$

But this is again a hypergeometric equation whose parameters α, β, and γ are now $-(1/2)n$, $(1/2)(n+1)$, and $1/2$, respectively. And since γ is not an integer, we can conclude from the results of the preceding section that this equation has two linearly independent solutions w_1 and w_2 about $t=0$ which are given by equations (7–35) and (7–39), respectively. Thus,

$$w_1 = F\left(-\frac{n}{2}, \frac{n+1}{2}; \frac{1}{2}; z^2\right) \qquad (7–60)$$

$$w_2 = zF\left(-\frac{n+1}{2}, \frac{n+2}{2}; \frac{3}{2}; z^2\right) \qquad (7–61)$$

They converge at least within the circle $|z| < 1$. When n is a nonnegative integer, the series for w_1 will terminate for even values of n and that for w_2 will terminate for odd values of n. However, the polynomials obtained in this manner must be identical, to within a constant factor, with the Legendre polynomials $P_n(z)$.

The associated Legendre equation

$$(1-z^2)w'' - 2zw' + \left[n(n+1) - \frac{m^2}{1-z^2}\right]w = 0 \qquad (7–62)$$

reduces to Legendre's equation when $m=0$. It is a Riemann-Papperitz equation for all values of m. Its P-symbol is

$$w = P\begin{pmatrix} 1 & -1 & \infty & \\ (1/2)m & (1/2)m & -n & z \\ -(1/2)m & -(1/2)m & n+1 & \end{pmatrix}$$

But upon transforming this to normal form, we find that

$$w= (z^2 -1)^{m/2}P\begin{pmatrix} 0 & 0 & \infty & \\ 0 & 0 & m-n & (1-z)/2 \\ -m & -m & m+n+1 & \end{pmatrix}$$

If we let u be the transformed dependent variable, then

$$w= (z^2 -1)^{m/2}u \tag{7-63}$$

and

$$u=P\begin{pmatrix} 0 & 0 & \infty & \\ 0 & 0 & m-n & (1-z)/2 \\ -m & -m & m+n+1 & \end{pmatrix}$$

$$=P\begin{pmatrix} 1 & -1 & \infty & \\ 0 & 0 & m-n & z \\ -m & -m & m+n+1 & \end{pmatrix}$$

Hence, u satisfies the equation

$$(1- z^2)\, u'' -2\,(m+1)\, zu' + (n-m)\,(n+m+1)\, u=0 \tag{7-64}$$

But when Legendre's equation (7–58) is differentiated m times, we obtain

$$(1-z^2)\frac{d^{m+2}w}{dz^{m+2}} -2\,(m+1)\,z\frac{d^{m+1}w}{dz^{m+1}} + (n-m)\,(n+m+1)\frac{d^m w}{dz^m} =0$$

Since P_n and Q_n are two linearly independent solutions of equation (7–58) about $z=1$, we see upon comparing equations (7–63), (7–64), and (7–65) that

$$P_n^m(z)= (z^2 -1)^{m/2}\frac{d^m}{dz^m}\,[P_n(z)]$$

and

$$Q_n^m(z)= (z^2 -1)^{m/2}\frac{d^m}{dz^m}\,[Q_n(z)]$$

are two linearly independent solutions to equation (7–62) about $z=1$. They are called the *associated Legendre functions* of the first and second kind, respectively.

CHAPTER 8

Confluent Hypergeometric Equation and Confluence of Singularities in Riemann-Papperitz Equation

Having studied equations whose only singular points are regular singular points, it is natural to approach the study of equations with irregular singular points by applying a limiting process wherein two or more regular singular points of an equation are allowed to approach one another. The process, whereby two or more singular points in any linear differential equation are allowed to come together in such a way that at least one of the corresponding exponents becomes infinite, is called *confluence*, provided the limiting form of the differential equation exists.

8.1 CONFLUENCE OF SINGULARITIES IN RIEMANN-PAPPERITZ EQUATION

We shall apply this process to the Riemann-Papperitz equation whose singular points are located at 0, b, and ∞ by letting the regular singular point b approach the regular singular point at ∞. Now if we are going to allow the exponents at the points b and ∞ to become infinite, we must prescribe the manner in which they approach infinity. This can be done by letting these exponents be functions of b whose values approach infinity as $b \to \infty$. If we suppose that the exponents are polynomials in b, it is not hard to show that the requirement that the limiting form of the differential equation exist implies that the exponents be, at most, linear in b. Now equation (7–8) is the general Riemann-Papperitz equation with one of its singular points at ∞. If we require that the two singular points in the finite part of the plane are at 0 and b and if we suppose the exponents at b are $\beta_1' + b\beta_2'$ and $\beta_1'' + b\beta_2''$ and those at ∞ are $\gamma_1' + b\gamma_2'$ and $\gamma_1'' + b\gamma_2''$, this equation becomes

$$\frac{d^2w}{dz^2} + \left[\frac{1 - \alpha' - \alpha''}{z} + \frac{1 - \beta_1' - \beta_1'' - b(\beta_2' + \beta_2'')}{z - b} \right] \frac{dw}{dz}$$

$$+ \left[\frac{\alpha'\alpha''b}{z} + \frac{(\beta_1' + b\beta_2')(\beta_1'' + b\beta_2'')b}{b - z} - (\gamma_1' + b\gamma_2')(\gamma_1'' + b\gamma_2'') \right] \frac{w}{(b - z)z} = 0$$

$$(8\text{--}1)$$

But if the coefficient of w is to remain finite as $b \to \infty$, we must require that

$$\beta_2'\beta_2'' - \gamma_2'\gamma_2'' = 0 \qquad (8\text{--}2)$$

Then upon taking the limit in equation (8–1), we obtain the equation

$$\frac{d^2w}{dz^2} + \left(\beta_2' + \beta_2'' + \frac{1 - \alpha' - \alpha''}{z} \right) \frac{dw}{dz}$$

$$+ \left[\frac{\alpha'\alpha''}{z^2} + \beta_2'\beta_2'' - \frac{(\alpha' - 1)\beta_2'' + \alpha''\beta_2' + \alpha_1}{z} \right] w = 0 \qquad (8\text{--}3)$$

where we have put

$$\alpha_1 = \gamma_1'\gamma_2'' + \gamma_1''\gamma_2' + (1 - \alpha' - \beta_1')\beta_2'' - (\alpha'' + \beta_1'')\beta_2'$$

This equation has a regular singular point at $z = 0$ and an irregular singular point at $z = \infty$.

8.2 **TRANSFORMATION TO KUMMER'S CONFLUENT HYPERGEOMETRIC EQUATION**

Making the change of dependent variable

$$w = z^{\alpha''} e^{-\beta_2'' z} u \tag{8-4}$$

transforms equation (8–3) into the equation

$$z \frac{d^2 u}{dz^2} + [\gamma - (\beta_2'' - \beta_2') z] \frac{du}{dz} - \alpha_1 u = 0$$

where we have put

$$\gamma = 1 + \alpha'' - \alpha' \tag{8-5}$$

And, if[91] $\beta_2'' \neq \beta_2'$, the change of independent variable

$$t = (\beta_2'' - \beta_2') z \tag{8-6}$$

leads to *Kummer's confluent hypergeometric equation*

$$t \frac{d^2 u}{dt^2} + (\gamma - t) \frac{du}{dt} - \alpha u = 0 \tag{8-7}$$

where $\alpha = \alpha_1 / (\beta_2'' - \beta_2')$.

Thus, once the solution to equation (8–7) is known, we can find the solutions to any equation of the form (8–3).

8.3 **SOLUTIONS TO KUMMER'S EQUATION: CONFLUENT HYPERGEOMETRIC FUNCTIONS**

We shall, therefore, seek a solution to Kummer's equation about the regular singular point $t = 0$. Just as in the case of the hypergeometric equation of

---

[91] If $\beta_2'' = \beta_2'$, the change of independent variable $z = (\tau^2/16\alpha_1)$ and the change of dependent variable $u = e^{-(1/2)\tau} v$ lead to the equation $\tau(d^2 v/d\tau^2) + (2\gamma - 1 - \tau) (dv/d\tau) - [(2\gamma - 1)/2] v = 0$, which is a special case of eq. (8–7).

Gauss, the characteristic exponents at this point are 0 and $1-\gamma$. We again suppose first that γ is neither zero nor a negative integer. Then equation (8–7) has a solution of the form $u_1 = \sum\limits_{n=0}^{\infty} a_n t^n$, which corresponds to the exponent $\rho = 0$. And we find, in the usual way, that

$$\sum_{n=0}^{\infty} a_n (n-1+\gamma) n t^{n-1} - \sum_{n=1}^{\infty} a_{n-1}(n-1+\alpha) t^{n-1} = 0$$

Therefore, a_0 is arbitrary and the recurrence relation is

$$a_n = \frac{(n-1+\alpha)}{n(n-1+\gamma)} a_{n-1} \qquad \text{for } n = 1, 2, \ldots$$

which can be solved to obtain

$$a_n = \frac{(\alpha)_n}{n!(\gamma)_n} a_0 \qquad \text{for } n = 1, 2, \ldots$$

Hence, by using the generalized hypergeometric notation, the solution u_1 can be expressed in the form

$$u_1 = a_0 \, {}_1F_1(\alpha; \gamma; t) \tag{8–8}$$

where the function

$${}_1F_1(\alpha; \gamma; t) = \sum_{n=0}^{\infty} \frac{(\alpha)_n}{n!(\gamma)_n} t^n \tag{8–9}$$

is called the *Pochhammer-Barnes confluent hypergeometric function*. Since the only singularities of equation (8–7) are at $t=0$ and $t=\infty$. the solution (8–9) must be an entire function.

Other commonly used notations for ${}_1F_1(\alpha; \gamma; t)$ are $M(\alpha; \gamma; t)$ and $\Phi(\alpha; \gamma; t)$. A complete discussion of confluent hypergeometric functions is given in reference 28.

Notice that by putting $t = z/\beta$ in the hypergeometric equation of Gauss (eq. (7–36)), we obtain the equation

$$z \left(1 - \frac{z}{\beta}\right) \frac{d^2 u}{dz^2} + \left[\gamma - (\alpha + \beta + 1) \frac{z}{\beta}\right] \frac{du}{dz} - \alpha u = 0$$

And upon taking the limit $\beta \to \infty$ in this equation, we arrive at the confluent hypergeometric equation (8–7). On the other hand, by putting $t = z/\beta$ in the hypergeometric function of Gauss (defined in eq. (7–33)), we obtain

$$F \left(\alpha, \beta; \gamma; \frac{z}{\beta}\right) = \sum_{n=0}^{\infty} \frac{(\alpha)_n (\beta)_n}{n! (\gamma)_n \beta^n} z^n$$

And since

$$\lim_{\beta \to \infty} \frac{(\beta)_n}{\beta^n} = 1$$

we see that at least formally

$$\lim_{\beta \to \infty} F \left(\alpha, \beta; \gamma; \frac{z}{\beta}\right) = {}_1F_1(\alpha; \gamma; z)$$

In a similar way, if γ is not a positive integer, consideration of the series corresponding to the exponent $1 - \gamma$ shows that

$$u_2 = t^{1-\gamma} \, {}_1F_1(\alpha + 1 - \gamma; 2 - \gamma; t) \tag{8–10}$$

is also a solution to equation (8–12)

Thus, if γ is not an integer, equations (8–9) and (8–10) provide a fundamental set of solutions to equation (8–7). But if γ is equal to 1, these two solutions are identical. And if γ is an integer other than 1, one of the hypergeometric functions in equations (8–9) and (8–10) is undefined and the other one provides a solution to the differential equation. However, when γ is a positive integer and α is not a positive integer smaller than γ, the other linearly independent solution is given by

$$u = {}_1F_1(\alpha; \gamma; t) \ln t - \sum_{n=0}^{\gamma-2} \frac{n!(1-\gamma)_{n+1}}{(1-\alpha)_{n+1}} t^{-(n+1)}$$

$$+ \sum_{n=1}^{\infty} \frac{(\alpha)_n}{(\gamma)_n n!} [H(\alpha, n) - H(\gamma, n) - H(1, n)] t^n$$

where the finite sum is to be omitted when $\gamma = 1$ and $H(\alpha, n)$ is defined in section 7.6. A corresponding result holds when γ is a nonpositive integer, and the remaining cases should be treated individually by the methods of chapter 6.

8.4 WHITTAKER'S FORM OF HYPERGEOMETRIC EQUATION

There is another standard form into which equation (8–7) (and, therefore, as a consequence, eq. (8–3)) can be transformed. To obtain this equation, we make the change of variable

$$W = t^{(1/2)\gamma} e^{-(1/2)t} u \tag{8–11}$$

in equation (8–7) to obtain the equation

$$\frac{d^2W}{dt^2} + \left[\frac{1}{t} \left(\frac{1}{2} \gamma - \alpha \right) - \frac{1}{4} + \frac{\gamma\left(1 - \frac{1}{2}\gamma\right)}{2t^2} \right] W = 0 \tag{8–12}$$

In order to introduce standard notation, we put

$$k = \frac{1}{2} \gamma - \alpha \tag{8–13}$$

and

$$m = \frac{1}{2} (\gamma - 1) \tag{8–14}$$

to get *Whittaker's confluent hypergeometric equation*

$$\frac{d^2W}{dt^2} + \left(-\frac{1}{4} + \frac{k}{t} + \frac{\frac{1}{4} - m^2}{t^2}\right)W = 0 \qquad (8\text{-}15)$$

It follows from equations (8–8), (8–10), (8–11), (8–13), and (8–14) that if $2m$ is not an integer

$$M_{k,\,m}(t) = t^{m+(1/2)}e^{-(1/2)t}\, {}_1F_1\left(\frac{1}{2} + m - k;\; 2m + 1;\; t\right) \qquad (8\text{-}16)$$

and $M_{k,\,-m}(t)$ are a fundamental set of solutions to equation (8–15) in the domain $0 < |t| < \infty$.

The notation $M_{k,\,m}$ is used in reference 25. The Whittaker's function $W_{k,\,m}$ is defined by

$$W_{k,\,m}(t) = \frac{\Gamma(-2m)}{\Gamma\left(-m-k+\frac{1}{2}\right)}M_{k,\,m}(t) + \frac{\Gamma(2m)}{\Gamma\left(m-k+\frac{1}{2}\right)}M_{k,\,-m}(t)$$

Since Whittaker's equation (8–15) is unaltered if we replace k by $-k$ and t and $-t$, it follows that

$$W = M_{-k,\,m}(-t) \qquad \text{and} \qquad W = M_{-k,\,-m}(-t) \qquad (8\text{-}17)$$

are also solutions to equation (8–15) in the region $0 < |t| < \infty$. And since this equation can have, at most, two linearly independent solutions, there must be constants C_1 and C_2, such that

$$M_{-k,\,m}(-t) = C_1 M_{k,\,m}(t) + C_2 M_{k,\,-m}(t) \qquad (8\text{-}18)$$

provided $2m$ is not an integer. But equations (8–9) and (8–16) show that $M_{k,\,m}(t)$ and $M_{-k,\,m}(-t)$ behave like $t^{m+(1/2)}$, that $M_{k,\,-m}$ behaves like $t^{-m+(1/2)}$ in the neighborhood of $t = 0$, and that

$$\lim_{t \to 0} \frac{1}{(-t)^{m+(1/2)}} M_{-k, m}(t) = \lim_{t \to 0} \frac{1}{t^{m+(1/2)}} M_{k, m}(t) = 1$$

Hence, equation (8–18) can hold only if $C_2 = 0$ and $C_1 = (-1)^{m+(1/2)}$. Therefore,

$$M_{-k, m}(-t) = (-1)^{m+(1/2)} M_{k, m}(t)$$

But substituting equation (8–16) in this equation shows that

$$e^t \, {}_1F_1\left(\frac{1}{2} + k + m; \, 1 + 2m; \, -t\right) = {}_1F_1\left(\frac{1}{2} - k + m; \, 1 + 2m; \, t\right)$$

And upon using equations (8–13) and (8–14) to express k and m in terms of α and γ, we obtain *Kummer's first formula*

$$\,{}_1F_1(\alpha; \, \gamma; \, t) = e^t \, {}_1F_1(\gamma - \alpha; \, \gamma; \, -t) \tag{8–19}$$

This equation holds even when γ is a positive integer, for in this case, we could have replaced $M_{k, -m}$ by the appropriate logarithmic solution and carried through the proof in the same way.

8.5 LAGUERRE POLYNOMIALS

If the parameter α in equation (8–7) is zero or a negative integer, the solution ${}_1F_1(\alpha; \, \gamma; \, t)$ (defined in eq. (8–9)) terminates after a finite number of terms and becomes a polynomial. This polynomial differs from the *generalized Laguerre polynomial*

$$L_n^{(b)}(t) = \frac{(1+b)_n}{n!} \, {}_1F_1(-n; \, 1+b; \, t) \qquad \text{for } n = 0, 1, 2, \ldots \tag{8–20}$$

only by a normalizing factor. When $b = 0$ in equation (8–20), we obtain the *simple Laguerre polynomial* L_n defined by

$$L_n(t) = {}_1F_1(-n; \, 1; \, t)$$

Equations (8–9) and (8–20) show that

$$L_n^{(b)}(t) = \sum_{j=0}^{n} \frac{(-n)_j (1+b)_n t^j}{j!(1+b)_j n!}$$

And since $(-n)_j/n! = (-1)^j/(n-j)!$

$$L_n^{(b)}(t) = \sum_{j=0}^{n} \frac{(1+b)_n (-t)^j}{(1+b)_j j!(n-j)!} \tag{8-21}$$

and

$$L_n(t) = \sum_{j=0}^{n} \frac{n!(-t)^j}{(j!)^2 (n-j)!} \tag{8-22}$$

8.6 BESSEL'S EQUATION

An important special case of equation (8–3) occurs when

$$
\left.
\begin{array}{l}
\alpha'' = -\alpha' = p \\[2mm]
\beta_2'' = -\beta_2' = i \\[2mm]
\alpha_1 = 2i\left(\alpha'' + \dfrac{1}{2}\right) = \left(p + \dfrac{1}{2}\right)2i
\end{array}
\right\} \tag{8-23}
$$

The equation

$$z^2 w'' + z w' + (z^2 - p^2) w = 0 \tag{8-24}$$

obtained in this manner is called *Bessel's equation*. When p is not a negative integer, the methods of chapter 6 can be applied in the usual way to show that (see section 7.5)

$$\sum_{k=0}^{\infty} \frac{(-1)^k z^{2k+p}}{2^{2k} k!(1+p)_k} = z^p {}_0F_1\left(-\,;\ 1+p;\ -\frac{z^2}{4}\right) \tag{8-25}$$

is a solution to equation (8–24) about the regular singular point $z = 0$. However, it is conventional to multiply this solution by the normalization factor $[2^p\Gamma(1+p)]^{-1}$ and to use equation (5–35) to eliminate the generalized factorial function in the denominator. When this is done, we obtain the suitable normalized solution

$$w_1 = J_p \qquad\qquad (8\text{–}26)$$

where

$$J_p(z) = \sum_{k=0}^{\infty} \frac{(-1)^k}{k!\,\Gamma(1+k+p)}\left(\frac{z}{2}\right)^{2k+p} \qquad\qquad (8\text{–}27)$$

is the *Bessel's function of the first kind* of order p. Notice that this function is defined even when p is a negative integer. And since $z = 0$ is the only finite singular point of the differential equation, the series must converge in the entire plane. Although Bessell functions were used by both Leonard Euler and Daniel Bernoulli before Bessel was born, the German astronomer Fredrick Wilhelm Bessel was the first to make a detailed study of them.

Since the function in equation (8–25) differs from J_p only by the normalization factor $[2^p\Gamma(1+p)]^{-1}$ when p is not a negative integer, it follows that

$$J_p(z) = \frac{\left(\dfrac{z}{2}\right)^p}{\Gamma(1+p)}\,{}_0F_1\left(-;\,1+p:-\frac{z^2}{4}\right) \qquad \text{for } p \neq -1, -2, \ldots \quad (8\text{–}28)$$

On the other hand, we know that equation (8–24) can be transformed into Kummer's confluent hypergeometric equation (8–7) by using the change of variables given by equations (8–4) to (8–6).

The solution J_p can, therefore, be expressed in terms of a linear combination of solutions of equation (8–7). Thus, if p is not an integer, it follows from equations (8–4) to (8–6), (8–8), (8–10), and (8–23) that

$$J_p(z) = Az^p e^{-iz}\,{}_1F_1\left(p+\frac{1}{2},\,1+2p;\,2iz\right) + Bz^{-p}e^{-iz}\,{}_1F_1\left(\frac{1}{2}-p;\,1-2p;\,2iz\right)$$

But upon comparing the behavior of these functions at $z=0$, we find, when p is not an integer, that this equation can only hold if $B=0$ and that $A=[2^p\Gamma(1+p)]^{-1}$. Hence,

$$J_p(z) = \frac{z^p e^{-iz}}{2^p \Gamma(1+p)} \, {}_1F_1\left(p+\frac{1}{2};\, 1+2p;\, 2iz\right)$$

Comparing this with equation (8–28) shows that

$${}_0F_1\left(-;\, 1+p;\, -\frac{z^2}{4}\right) = e^{-iz} \, {}_1F_1\left(p+\frac{1}{2};\, 1+2p;\, 2iz\right)$$

Then by putting $\zeta = iz$ and $\alpha = p+(1/2)$, we obtain *Kummer's second formula*

$${}_0F_1\left(-;\, \alpha+\frac{1}{2};\, \frac{1}{4}\zeta^2\right) = e^{-\zeta} \, {}_1F_1\left(\alpha;\, 2\alpha;\, 2\zeta\right)$$

When p is a negative integer, there is a close relationship between J_p and J_{-p}. Thus, suppose $p=-n$ for $n=1, 2, \ldots$. Then since the Γ-function has poles at each of the nonpositive integers, all the terms in the sum (8–27) will vanish for $k < n$; and we obtain

$$J_{-n}(z) = \sum_{k=n}^{\infty} \frac{(1)^k}{k!\,\Gamma(1+k-n)}\left(\frac{z}{2}\right)^{2k-n}$$

But this becomes, upon shift of index from k to $k-n$,

$$J_{-n}(z) = \sum_{k=0}^{\infty} \frac{(-1)^{k+n}}{(k+n)!\,\Gamma(1+k)}\left(\frac{z}{2}\right)^{2k+n}$$

Hence, it follows from equation (5–36) that

$$J_{-n}(z) = (-1)^n J_n(z) \qquad \text{for } n=1, 2, \ldots \qquad (8\text{–}29)$$

243

Since $w_1 = J_p$ is a solution to equation (8–29) for all values of p and since the parameter p enters equation (8–24) only as p^2, the function

$$w_2 = J_{-p}(z) \tag{8–30}$$

must also be a solution to this equation for all values of p. When p is an integer, equation (8–29) shows that the solutions w_1 and w_2 are not linearly independent. However, for any nonintegral value of p, consideration of the behavior of the functions J_p and J_{-p} in the neighborhood of $z=0$ shows that w_1 and w_2 constitute a fundamental set of solutions. In order to obtain a second linearly independent solution to equation (8–24) when p is an integer, notice that for any value of p

$$\left.\begin{array}{l} z^2 J_p'' + z J_p' + (z^2 - p^2) J_p = 0 \\[2mm] z^2 J_{-p}'' + z J_{-p}' + (z^2 - p^2) J_{-p} = 0 \end{array}\right\} \tag{8–31}$$

Since the series (8–27) is uniformly convergent, it can be shown that J_p, J_{-p}, and their derivatives with respect to z are all differentiable with respect to the parameter p. Hence, upon differentiating equations (8–31) with respect to p and subtracting the result, we obtain

$$z^2 \mathscr{J}_p'' + z \mathscr{J}_p' + (z^2 - p^2) \mathscr{J}_p = 2p[J_p - (-1)^n J_{-p}] \tag{8–32}$$

where we have put

$$\mathscr{J}_p \equiv \frac{\partial J_p}{\partial p} - (-1)^n \frac{\partial J_{-p}}{\partial p}$$

By taking the limit as $p \to n$ in equation (8–32) and using equation (8–29), we find that the *Bessel function of the second kind*

$$Y_n(z) \equiv \frac{1}{\pi} \lim_{p \to n} \left[\frac{\partial J_p(z)}{\partial p} - (-1)^n \frac{\partial J_{-p}(z)}{\partial p} \right] \tag{8–33}$$

is a solution to equation (8–24) with $p = n$. But upon substituting equation (8–27) into this expression and noting that

$$\lim_{z \to -k} \frac{1}{\Gamma^2(z)} \frac{d\Gamma}{dz} = (-1)^{k+1} k! \qquad \text{for } k = 0, 1, 2, \ldots$$

it follows after some algebraic manipulation, which we shall omit here, that

$$Y_n(z) = \frac{2}{\pi} \left(\gamma + \ln \frac{z}{2} \right) J_n(z) - \frac{1}{\pi} \sum_{k=0}^{n-1} \frac{(h-k-1)!}{k!} \left(\frac{z}{2} \right)^{2k-n}$$

$$- \frac{1}{\pi} \sum_{k=0}^{\infty} \frac{(-1)^k}{k!(n+k)!} (H_k + H_{n+k}) \left(\frac{z}{2} \right)^{2k+n} \qquad (8\text{--}34)$$

where the finite sum is to be omitted if $n = 0$, H_k is defined by equation (6–89), and γ is *Euler's constant* defined by

$$\gamma = \lim_{k \to \infty} (H_k - \ln k) = 0.5772156649 \ldots \qquad (8\text{--}35)$$

It is easy to see from equations (8–27) and (8–34) that Y_n and J_n are a fundamental set of solutions.

It is convenient to extend the definition (8–33) of Y_n to nonintegral values of p. This is now done by putting

$$Y_p(z) = \frac{J_p(z) \cos p\pi - J_{-p}(z)}{\sin p\pi} \qquad (8\text{--}36)$$

when p is not an integer. Then Y_p and J_p are a fundamental set of solutions. And both the numerator and denominator of this expression vanish when p is an integer. But by using L'Hospital's rule, we find that for $n = 0, 1, 2, \ldots$

$$\lim_{p \to n} Y_p(z) = \lim_{p \to n} \frac{J_p \cos p\pi - J_{-p}}{\sin p\pi} = \lim_{p \to n} \frac{\dfrac{\partial J_p}{\partial p} \cos \pi p - \dfrac{\partial J_{-p}}{\partial p} - \pi J_p \sin p\pi}{\pi \cos p\pi}$$

$$= \frac{1}{\pi} \lim_{p \to n} \left(\frac{\partial J_p}{\partial p} - \frac{1}{\cos p\pi} \frac{\partial J_{-p}}{\partial p} \right) = \frac{1}{\pi} \lim_{p \to n} \left[\frac{\partial J_p}{\partial p} - (-1)^n \frac{\partial J_{-p}}{\partial p} \right]$$

And comparing this with equation (8–33) shows that definition (8–36) is indeed an extension of definition (8–33).

For large values of z with $|\arg z| < \pi$, the Bessel functions J_p and Y_p behave like

$$\left.\begin{array}{l} J_p(z) \sim \left(\dfrac{2}{\pi z} \right)^{1/2} \cos\left(z - \dfrac{p\pi}{2} - \dfrac{\pi}{4} \right) \\[3mm] Y_p(z) \sim \left(\dfrac{2}{\pi z} \right)^{1/2} \sin\left(z - \dfrac{p\pi}{2} - \dfrac{\pi}{4} \right) \end{array}\right\} \qquad (8\text{--}37)$$

However, it is sometimes convenient to have solutions to Bessel's equation which tend to zero exponentially as $|z| \to \infty$ in the entire half plane $\mathscr{Im}\, z > 0$ or in the half plane $\mathscr{Im}\, z < 0$. But since equations (8–37) show that for large z

$$J_p(z) \pm iY_p(z) \sim \left(\frac{2}{\pi z} \right)^{1/2} e^{\pm i[z - (p\pi/2) - (\pi/4)]}$$

we see this property is possessed by functions $J_n \pm iY_p$. For this reason, the *Hankel functions of the first and second kind* $H_p^{(1)}$ and $H_p^{(2)}$, respectively, are defined by

$$H_p^{(1)}(z) = J_p(z) + iY_p(z)$$

$$H_p^{(2)}(z) = J_p(z) - iY_p(z)$$

It is not hard to show that these solutions are a fundamental set.

In a number of applications, we encounter the equation

$$z^2 w'' + zw' - (z^2 + p^2)w = 0 \qquad (8\text{--}38)$$

which is obtained from Bessel's equation by replacing z by iz. Although $J_p(iz)$ and $Y_p(iz)$ are a fundamental set of solutions of equation (8–38), it is convenient to introduce new functions which are real for real values of z (at least for real values of p). To this end the modified *Bessel functions of the first and third kind* I_p and K_p, respectively, are defined by

$$I_p(z) = e^{-i(p\pi/2)} J_p(iz)$$

$$K_p(z) = \frac{\pi i}{2} e^{i(p\pi/2)} H_p^{(1)}(iz) = \frac{\pi i}{2} e^{i(p\pi/2)} [J_p(iz) + iY_p(iz)]$$

There are a number of useful relations between the Bessel functions and their derivatives. Thus, it follows from equation (8–27) and equation (5–31) that

$$z^p J_p(z) = \sum_{k=0}^{\infty} \frac{(-1)^k z^{2k+2p}}{2^{2k+p} k! (k+p) \Gamma(k+p)}$$

But upon differentiating both sides with respect to z, we find that

$$\frac{d}{dz}(z^p J_p(z)) = \sum_{k=0}^{\infty} \frac{(-1)^k z^{2k+2p-1}}{2^{2k+p+1} k! \Gamma(k+p)} = z^p J_{p-1}(z) \tag{8–39}$$

In a similar way it can be shown that

$$\frac{d}{dz} z^{-p} J_p(z) = -z^{-p} J_{p+1}(z) \tag{8–40}$$

After carrying out the differentiations in equations (8–39) and (8–40), we obtain

$$z \frac{dJ_p}{dz} = z J_{p-1} - p J_p \tag{8–41}$$

$$z \frac{dJ_p}{dz} = -z J_{p+1} + p J_p \tag{8–42}$$

And by adding and subtracting these equations, we find

$$2 \frac{dJ_p}{dz} = J_{p-1} - J_{p+1} \tag{8–43}$$

and

$$2p J_p = z J_{p-1} + z J_{p+1} \tag{8–44}$$

The formulas (8–39) to (8–44) are very useful when dealing with Bessel functions. They also apply to the Bessel functions Y_p, $H_p^{(1)}$, and $H_p^{(2)}$. However, they must be modified slightly for the functions I_p and K_p (due to the presence of iz).

A much more complete treatment of Bessel functions can be found in reference 29.

8.7 WEBER'S EQUATION

Another equation, which has been extensively studied, can be obtained from equation (8–3) by putting

$$\alpha' = \frac{1}{2} \qquad \alpha'' = 0 \qquad \beta_2'' = -\beta_2' = \frac{1}{4} \qquad \alpha_1 = -\frac{n}{4} \quad (8\text{–}45)$$

to get

$$zw'' + \frac{1}{2} w' + \frac{1}{4}\left(\frac{1}{2} + n - \frac{1}{4} z\right) w = 0 \qquad (8\text{–}46)$$

and then making the change of independent variable

$$z = \zeta^2 \qquad (8\text{–}47)$$

to obtain *Weber's equation*

$$\frac{d^2 w}{d\zeta^2} + \left(n + \frac{1}{2} - \frac{1}{4}\zeta^2\right) w = 0 \qquad (8\text{–}48)$$

This equation was first studied by Hermite (1864) and then by Weber. Its solutions are called *parabolic cylinder functions* or *Weber functions*.

Since equation (8–46) can be transformed into the confluent hypergeometric equation (8–7) by the change of variables (8–4) and (8–6), equation (8–48) can be transformed into equation (8–7) by the change of variables (8–4) to (8–6) and (8–47). Hence, the solutions to equation (8–48) can be expressed in terms of the solutions to equation (8–7). Therefore, use of equations (8–4) to (8–6), (8–8), (8–10), (8–45), and (8–47) shows that a fundamental set of solutions to Weber's equation (w_1, w_2) is given by

$$w_1 = e^{-(1/4)\zeta^2} {}_1F_1\left(-\frac{n}{2}; \frac{1}{2}; \frac{1}{2}\zeta^2\right) \qquad (8\text{–}49)$$

$$w_2 = e^{-(1/4)\zeta^2} \zeta^2 {}_1F_1\left(-\frac{n-1}{2}; \frac{3}{2}; \frac{1}{2}\zeta^2\right) \qquad (8\text{–}50)$$

Notice that if n is a positive even integer, the hypergeometric function in equation (8–49) terminates and becomes a generalized Laguerre polynomial; while if n is a positive odd integer, the hypergeometric function in equation (8–50) becomes a generalized Laguerre polynomial. Thus, it follows from equations (8–20), (8–49), and (8–50) that, if n is a positive integer, equation (8–48) has one solution w_3 of the form

$$w_3 = e^{-(1/4)\zeta^2} H_n(\zeta) \qquad \text{for } n = 1, 2, \ldots \qquad (8\text{–}51)$$

where H_n is the nth *Hermite polynomial*, which is defined in terms of the generalized Laguerre polynomials by

$$\left. \begin{aligned} H_{2k}(\zeta) &= (-1)^k 2^{2k} k! L_k^{(-1/2)}(\zeta^2) \\ H_{2k+1}(\zeta) &= (-1)^k 2^{2k+1} k! \zeta L_k^{-(1/2)}(\zeta^2) \end{aligned} \right\} k = 0, 1, 2, \ldots \qquad (8\text{–}52)$$

8.8 OTHER EQUATIONS STUDIED

Although we have been able to give a complete presentation of the general theory of the solutions to second-order linear equations about regular singular points, apart from relatively trivial cases, only the solutions to those equations which can be transformed into the hypergeometric equations of Gauss and of Kummer have been exhaustively studied. The equations of *Lamé* and *Mathieu* have also been studied, but only a small amount of knowledge of the functions defined by these equations has been obtained. The *Mathieu equation* is

$$\frac{d^2w}{dz^2} + (a + k^2 \cos^2 z)w = 0$$

where a and k are parameters, and the *Lame equation* is

$$\frac{d^2w}{dz^2} + [\eta + n(n+1)m\, sn^3(z|m)]w = 0$$

where $sn(z|m)$ is the sine amplitude elliptic function of modulus m, and η, n, and m are parameters. When $n=2$, its solutions can be expressed in terms of elliptic functions. For more information about these equations, the reader is referred to reference 3. An extremely complete discussion of Mathieu's equation with many applications is given in reference 30.

All the equations which have been discussed in the last two chapters can be derived from a single equation with six distinct regular singular points (with the characteristic exponents at each singular point differing by 1/2) by allowing these singular points to coalesce. The coalescence of any two such points is a regular singular point whose exponents differ by an arbitrary amount. The coalescence of three or more such points will result in an irregular singular point. The reader is referred to reference 3, chapter 20, for a more detailed study of these matters.

CHAPTER 9

Solutions Near Irregular Singular Points: Asymptotic Expansions

9.1 GENERAL CHARACTER OF SOLUTIONS

Let z_0 be an isolated singular point of the differential equation

$$\frac{d^2w}{dz^2} + p(z)\,\frac{dw}{dz} + q(z)w = 0 \tag{9-1}$$

We have shown in chapter 6 that it is always possible to construct a fundamental set of solutions to this equation on a punctured circular region about z_0 whenever z_0 is a regular singular point. And these solutions can be expressed in terms of certain convergent power series whose coefficients can be calculated successively. This occurs because the functions $f_1(z)$ and $f_2(z)$ in the canonical basis (6–20) can have, at most, poles at z_0. However, this will not be the case when the point z_0 is an irregular singular point of equation (9–1) since at least one of the two functions $f_1(z)$ and $f_2(z)$ will then have an essential singularity at z_0. This function will therefore have a series expansion about z_0 of the form (6–21) in which infinitely many negative powers of $z - z_0$ are present. And the coefficients of this series are determined by an infinite set of equations which cannot be solved recursively. For this reason we do not have a convenient method of finding the corresponding solution. In addition, the characteristic exponents are determined by a transcendental equation (which usually cannot be solved explicitly) instead of by an algebraic equation. Finally, the series once found are usually slowly convergent.

9.2 FORMAL POWER SERIES

Even if z_0 is an irregular singular point, it may still happen that the function $f_1(z)$ in the canonical basis (6–20) has, at most, a pole at z_0, in which case it would still be possible to recursively construct [92] one solution about this point. And this solution would be of the form[93]

$$w = (z - z_0)^\rho f(z) \tag{9–2}$$

where $f(z)$ is analytic in a neighborhood of z_0, and therefore has a convergent power series expansion

$$f(z) = \sum_{n=0}^{\infty} a_n (z - z_0)^n \tag{9–3}$$

about z_0. We assume, without loss of generality, that the exponent ρ is adjusted to make $a_0 \neq 0$ and therefore to make

$$f(z_0) \neq 0 \tag{9–4}$$

A second solution could then be found, at least in principle, from equation (6–63).

When $p(z)$ has an essential singularity at z_0, equation (9–1) can possess a solution of the form (9–2) only when $q(z)$ is equal to zero or has an essential singularity at z_0. But suppose that $p(z)$ has a pole at z_0, say of order m. Thus,

$$p(z) = (z - z_0)^{-m} P(z) \tag{9–5}$$

where $P(z)$ is analytic at $z = z_0$. Now if equation (9–1) possesses a regular solution of the form (9–2), this solution together with (9–5) can be substituted into equation (9–1) to obtain

---

[92] If it were the function $f_2(z)$ which had the pole at z_0 and the function $f_1(z)$ which had the essential singularity, then either we could (when $a = 0$) change the notation so that the function $f_1(z)$ had the pole at z_0 or (when $a \neq 0$) the solution w_2 would depend upon $f_1(z)$ (through $w_1(z)$) and could therefore not be constructed recursively.

[93] Recall (ch. 6) that a solution of this type is called a regular solution.

$$q(z) = -\frac{f''(z)}{f(z)} - \frac{2\rho}{z-z_0}\frac{f'(z)}{f(z)} - \frac{\rho(\rho-1)}{(z-z_0)^2}$$

$$-\left[\frac{1}{(z-z_0)^m}\frac{f'(z)}{f(z)} + \frac{\rho}{(z-z_0)^{m+1}}\right]P(z) \qquad (9\text{--}6)$$

And since $f(z_0) \neq 0$, it follows that the functions f''/f, f'/f, and P are analytic at $z = z_0$.

First, suppose that m is either zero or 1. Then equation (9–6) shows that $q(z)$ has, at most, a pole of order two. We therefore conclude that when $p(z)$ has, at most, a simple pole at z_0, equation (9–1) will possess a regular solution of the form (9–2) if, and only if, z_0 is a regular singular point.

Now, suppose that $m > 1$. Then equation (9–6) shows that $q(z)$ has, at most, a pole of order $m + 1$. Hence, we conclude that when $p(z)$ has a pole of order $m > 1$ at a point z_0, equation (9–1) can possess a regular solution about this point only if $q(z)$ has, at most, a pole of order $m + 1$ at z_0.

The corresponding result for the case where the point z_0 is at ∞ can be obtained in the usual way by applying these results at the origin of the ζ-plane to the coefficients of the transformed equation given by equations (6–28) and (6–29). This leads to the conclusion that if $p(z)$ has a pole of order[94] $m \geqslant 0$ about $z = \infty$, equation (9–1) possesses a solution about this point of the form

$$z^\rho \sum_{n=0}^{\infty} \frac{a_n}{z^n}$$

only if the highest power of z which occurs in the Laurent series expansion of $q(z)$ about $z = \infty$ is $m - 1$.

Now suppose that z_0 is a finite point, $p(z)$ has a pole of order m at z_0, and equation (9–1) possesses a solution of the form (9–2) about z_0. Then, $q(z)$ must be of the form

$$q(z) = (z-z_0)^{-m-1}Q(z) \qquad (9\text{--}7)$$

---

[94] Recall that we say that a function has a pole of order zero at a point if it is analytic at that point.

where $Q(z)$ is analytic at $z=z_0$. This function $Q(z)$ and the function $P(z)$ which appears in the representation (9–5) of $p(z)$ can be expanded about z_0 in the convergent power series

$$\left.\begin{array}{c} P(z) = \displaystyle\sum_{n=0}^{\infty} p_n(z-z_0)^n \\[3mm] Q(z) = \displaystyle\sum_{n=0}^{\infty} q_n(z-z_0)^n \end{array}\right\} \tag{9–8}$$

where $p_0 \neq 0$.

We can now proceed much as we did in chapter 6 for the case of a regular singular point by substituting equations (9–2), (9–3), (9–5), (9–7), and (9–8) with $m > 1$ into equation (9–1) and then collecting the coefficients of like powers of $z-z_0$. When this procedure is carried out, we find first that the indicial equation is not quadratic (as in the case of a regular singular point) but that it is an equation of the first degree which is

$$\rho p_0 + q_0 = 0 \tag{9–9}$$

This is consistent with the fact that, at an irregular singular point, at most one solution of the form (9–2) can occur.

Next, the coefficients of the series (9–3) can still be calculated recursively, but the resulting series will now either *terminate after a finite number of terms* or else it will *diverge* for all $|z-z_0| > 0$ (ref. 3, p. 421). Thus, a regular solution will only exist in the exceptional case when the series (9–3) contains a finite number of terms.

For example, the equation

$$w'' + 2\,\frac{(z-a)}{z^2}\,w' - \frac{1}{z^3}\left(a + \frac{3}{4}\,z\right) w = 0$$

has an irregular singular point at $z = 0$. And it has a series solution about this point given by

$$w = z^{-1/2}\left(1 - \frac{1}{2a}\,z\right)$$

which terminates after two terms.

Even though it is possible to calculate as many terms as desired when the series does not terminate, the solution obtained in this manner will not have any meaning in the usual sense since the series does not converge. This should not be taken to mean, however, that this "formal" solution is not useful.

9.3 ASYMPTOTIC CHARACTER OF SOLUTIONS

In order to introduce the concepts which allow the formal solutions obtained in the preceding section to be utilized, it is convenient to consider an example given by Euler in 1754. Thus, the differential equation

$$w'' + \frac{3z + 1}{z^2}\, w' + \frac{1}{z^2}\, w = 0 \qquad (9\text{--}10)$$

has an irregular singular point at the origin and a regular singular point at infinity. Since the order of the pole of the coefficient of w at $z = 0$ does not exceed the order of the pole of the coefficient of w' at $z = 0$ by more than 1, we know that equation (9–10) possesses one (and only one) formal solution about $z = 0$ of the form

$$w = z^\rho \sum_{n=0}^{\infty} a_n z^n \qquad (9\text{--}11)$$

We can therefore proceed just as in the case of a regular singular point and substitute this expansion into equation (9–10), shift the indices to the lowest one present, and collect terms to obtain

$$\sum_{n=1}^{\infty} (n + \rho)^2 a_{n-1} z^{n+\rho-1} + \sum_{n=0}^{\infty} (n + \rho) a_n z^{n+\rho-1} = 0$$

Then by equating the coefficients of the various powers of z to zero, we find that since $a_0 \neq 0$

$$\rho = 0 \qquad (9\text{--}12)$$

255

and

$$(n + \rho)^2 a_{n-1} = -(n + \rho) a_n \qquad \text{for } n = 1, 2, 3, \ldots \qquad (9\text{--}13)$$

The indicial equation (9–12) is consistent with the general indicial equation (9–9) since, in this case, $q_0 = 0$. When equation (9–12) is substituted into the recurrence relation (9–13), we obtain

$$a_n = -n a_{n-1} \qquad \text{for } n = 1, 2, 3, \ldots \qquad (9\text{--}14)$$

And by proceeding in the usual way, we conclude from this, that

$$a_n = (-1)^n n! a_0$$

Finally, substituting this, together with equation (9–12), into equation (9–11) shows that equation (9–10) possesses the formal solution

$$w = \sum_{n=0}^{\infty} (-1)^n n! z^n \qquad (9\text{--}15)$$

A simple application of the ratio test, however, shows that this series diverges for all $z \neq 0$. Hence, as we have already indicated in the general case, equation (9–15) is not a regular solution. However, we might anticipate that, since equation (9–15) was determined in a formal manner by the differential equation, it might still, in some sense, represent a solution to that differential equation.

In order to see what this is, note that equations (5–29) and (5–36) show that

$$n! = \int_0^{\infty} e^{-t} t^n dt \qquad (9\text{--}16)$$

However, when this is substituted into equation (9–15), we find that

$$w = \sum_{n=0}^{\infty} \int_0^{\infty} e^{-t} (-1)^n z^n t^n dt$$

And again, proceeding formally we interchange the order of summation and integration to obtain

$$w = \int_0^{\infty} e^{-t} \sum_{n=0}^{\infty} (-zt)^n dt$$

Finally, summing the geometric series shows that

$$w = \int_0^{\infty} \frac{e^{-t}}{1 + zt} \, dt \tag{9-17}$$

Now this integral converges uniformly [95] for all z in any region R bounded away from the negative real axis. In fact, differentiating with respect to z and interchanging the order of integration and differentiation shows that

$$\frac{dw}{dz} = -\int_0^{\infty} \frac{te^{-t}}{(1 + zt)^2} \, dt \tag{9-18}$$

---

[95] This means that for any positive number ϵ, no matter how small, there exists a positive number $T(\epsilon)$ which depends on ϵ *but not on* z such that $\left| \int_t^{\infty} [e^{-t}/(1 + zt)] \, dt \right| < \epsilon$ for all $t \geq T(\epsilon)$ and all z in R.

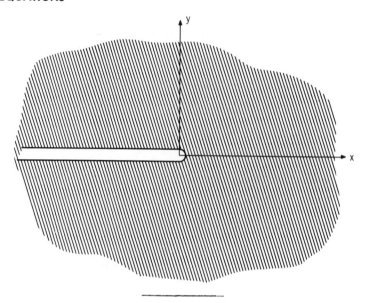

FIGURE 9–1.—Region of convergence of $\int_0^\infty \dfrac{te^{-t}}{(1+zt)^2}\,dt$ and $\int_0^\infty \dfrac{e^{-t}}{(1+zt)^2}\,dt$.

But since this integral converges uniformly for all values of z in any region bounded away from the negative real axis (fig. 9–1), we can conclude that the derivative exists and that interchanging the order of integration and differentiation was in fact justified, provided z is not on the negative real axis (ref. 31). Hence, $\int_0^\infty [e^{-t}/(1+zt)]\,dt$ is an analytic function of z everywhere in the z-plane cut along the negative real axis and is therefore infinitely differentiable in this region.

Upon multiplying equation (9–17) by z and equation (9–18) by z^2 and then adding the result, we get

$$z^2 w' + zw = - z^2 \int_0^\infty \frac{e^{-t}t}{(1+zt)^2}\,dt + z \int_0^\infty \frac{e^{-t}}{1+zt}\,dt$$

$$= \int_0^\infty e^{-t}\,\frac{z(1+zt)-z^2 t}{(1+zt)^2}\,dt = \int_0^\infty \frac{e^{-t}z}{(1+zt)^2}\,dt$$

And after integration by parts, this becomes

$$z^2 w' + zw = -\frac{e^{-t}}{1+zt}\Big|_0^\infty - \int_0^\infty \frac{e^{-t}}{1+zt}\,dt = 1 - w$$

which shows that w satisfies the equation

$$z^2 w' + (z+1)w = 1$$

provided z is not on the negative real axis. Since w is infinitely differentiable, we can certainly differentiate this equation with respect to z. But when this is done we obtain equation (9–10). Thus, the function w given by equation (9–17) provides a perfectly valid solution to equation (9–10). This solution was obtained by formally summing the divergent series (9–15), which was, in turn, obtained by formally solving equation (9–10). It is, therefore, natural to ask in which sense the divergent series (9–15) represents the integral (9–17) and, therefore, as a consequence, in what sense it represents a solution to equation (9–10). To answer this question, let

$$S_m(z) = \sum_{n=0}^{m} (-1)^n n!\, z^n \tag{9-19}$$

be the mth partial sum of the series (9–15). Now from elementary algebra, we know that the sum of the finite geometric series is

$$\sum_{n=0}^{m} r^n = \frac{1 - r^{m+1}}{1 - r}$$

which becomes, upon setting $r = -zt$,

$$\frac{1}{1+zt} = \frac{(-zt)^{m+1}}{1+zt} + \sum_{n=0}^{m} (-1)^n (zt)^n$$

But, upon multiplying both sides by e^{-t} and integrating between 0 and ∞, we find that

$$\int_0^\infty \frac{e^{-t}}{1+zt}\,dt = \int_0^\infty \sum_{n=0}^{m} (-1)^n (zt)^n e^{-t} dt + R_m(z)$$

where we have put

$$R_m(z) = (-z)^{m+1} \int_0^\infty \frac{t^{m+1}e^{-t}}{1+zt}\, dt \qquad (9-20)$$

And since it is always legitimate to interchange the order of integration and summation for a finite sum, this becomes

$$\int_0^\infty \frac{e^{-t}}{1+zt}\, dt - \sum_{n=0}^m (-1)^n z^n \int_0^\infty t^n e^{-t} dt = R_m(z)$$

Hence, substituting in equations (9–16) and (9–19) shows that

$$\int_0^\infty \frac{e^{-t}}{1+zt}\, dt - S_m(z) = R_m(z) \qquad (9-21)$$

Now since

$$|1 + zt| = |1 + xt + iyt| = \sqrt{(1+xt)^2 + y^2 t^2}$$

it follows that

$$|1 + zt| \geqslant \sqrt{(1+xt)^2} = |1 + xt| \geqslant 1 \qquad \text{for} \quad \mathscr{R}e\ z = x > 0$$

And for $\mathscr{R}e\ z = x \leqslant 0$ we find that

$$|1 + zt|\ \sqrt{(xt)^2 + (yt)^2} = \{(1+xt)^2(xt)^2 + y^2 t^2[(1+xt)^2 + (xt)^2] + (yt)^4\}^{1/2}$$

$$\geqslant \sqrt{y^2 t^2[(1+xt)^2 + (xt)^2]} \geqslant \frac{1}{\sqrt{2}}|yt|$$

Hence

$$|1 + zt| \geqslant \frac{1}{\sqrt{2}}\frac{|yt|}{\sqrt{(xt)^2 + (yt)^2}} = \frac{1}{\sqrt{2}}\frac{|y|}{\sqrt{x^2 + y^2}} = \frac{1}{\sqrt{2}}|\sin\phi| \qquad \text{for} \quad \mathscr{R}e\ z = x < 0$$

where ϕ is the argument of z. We have therefore shown that

$$\frac{1}{|1 + zt|} \leqslant 1 \qquad \text{for } \mathscr{R}e \ z \geqslant 0 \tag{9-22}$$

and

$$\frac{1}{|1 + zt|} \leqslant \sqrt{2} \ |\text{cosec } \phi| \qquad \text{for } \mathscr{R}e \ z \leqslant 0; \ |\phi| < \pi \tag{9-23}$$

It now follows from equations (9–20) and (9–22) that when $\mathscr{R}e \ z \geqslant 0$

$$|R_m(z)| = |z|^{m+1} \left| \int_0^\infty \frac{t^{m+1}e^{-t}}{1 + zt} \, dt \right|$$

$$\leqslant |z|^{m+1} \int_0^\infty \left| \frac{t^{m+1}e^{-t}}{1 + zt} \right| \, dt \leqslant |z|^{m+1} \int_0^\infty t^{m+1}e^{-t}dt$$

And equations (9–16) and (9–21) therefore show that

$$|w_1(z) - S_m(z)| \leqslant |z|^{m+1}(m+1)! \qquad \text{for } \mathscr{R}e \ z > 0 \tag{9-24}$$

where

$$w_1(z) \equiv \int_0^\infty \frac{e^{-t}}{1 + zt} \, dt \tag{9-25}$$

is the solution to equation (9–10). In a similar way, it follows from equation (9–23) that

$$|w_1(z) - S_m(z)| \leqslant \sqrt{2}| \text{ cosec } \phi| \ |z|^{m+1}(m+1)! \qquad \text{for } \mathscr{R}e \ z < 0; \ |\phi| < \pi$$

$$\tag{9-26}$$

Thus, in view of equations (9–15) and (9–19), the inequalities (9–24) and (9–26) show that, provided ϕ, the argument of z, is smaller in absolute value than π, the error incurred by computing the solution $w_1(z)$ by means of the divergent series (9–15) goes to zero rapidly as $z \to 0$. Thus, for any fixed m, any desired accuracy can be obtained by taking $|z|$ sufficiently near zero. In particular, the inequality (9–24) shows that, provided $\mathscr{R}e\ z > 0$, the error is never greater in magnitude than the first term omitted in the series.

Now recall that a series $\sum\limits_{n=0}^{\infty} a_n$ is said to converge to a sum A if its partial sums $t_m = \sum\limits_{n=0}^{m} a_n$ satisfy the relation

$$\lim_{m \to \infty} |A - t_m| = 0$$

The series (9–15) certainly does not converge to the solution $w_1(z)$ since

$$\lim_{m \to \infty} |w_1(z) - S_m(z)| \neq 0 \qquad (9\text{--}27)$$

But this series is said to be an asymptotic expansion (which will be defined subsequently) of $w_1(z)$ since, as can be seen from equations (9–24) and (9–26), it has the property that

$$\lim_{z \to 0} \frac{w_1(z) - S_m(z)}{z^m} = 0 \qquad \text{for } |\phi| \leq \pi + \epsilon \qquad (9\text{--}28)$$

for any $\epsilon > 0$.

9.4 ASYMPTOTIC EXPANSIONS

Roughly speaking, equation (9–28) shows that the difference between the solution $w_1(z)$ and the mth partial sum $S_m(z)$ of the series (9–15) approaches zero more rapidly than z^m as $z \to 0$. Comparing the definition of convergence with the corresponding condition (9–28) for a series to be asymptotic, shows that the essential difference between these two concepts can be stated in the following way: If the series is convergent to $w_1(z)$, then z is held fixed and the difference $w_1(z) - S_m(z)$ approaches zero as m approaches infinity. If the series is asymptotic to $w_1(z)$, then the number m of terms is held fixed and the difference $w_1(z) - S_m(z)$ approaches zero (at a specified rate) as z approaches some fixed value, say z_0.

There is no reason why a given series cannot be both asymptotic and convergent at the same time. For example, the geometric series $\sum_{n=0}^{\infty} z^n$ is convergent to $1/(1-z)$ for $|z| < 1$ and it is also an asymptotic expansion of this function as $z \to 0$.

When a series is convergent as well as asymptotic, the approximation can usually be improved at fixed z by taking more and more terms in the partial sum. However, it happens more frequently that an asymptotic series is divergent.[96] When this is the case, there is some limiting value to the accuracy of the approximation which can be obtained at each fixed value of z. This limiting value is attained when the optimal number of terms is retained in the expansion. Thus, for the expansion (9–15) with $z > 0$, the right side of the inequality (9–24) is a minimum when $m \approx |z|^{-1}$. For this value of m, $S_m(z)$ will provide the best approximation to $w_1(z)$.

The theory of asymptotic expansion was initiated by Stieltjes and by Poincare at the end of the nineteenth century. Before giving the general definition of an asymptotic expansion, it is convenient to introduce the concept of order. Thus, if ϕ and ψ are complex functions of z, we write

$$\phi = O(\psi) \qquad \text{as } z \to z_0$$

if there exists a finite constant A such that

$$\lim_{z \to z_0} \frac{\phi(z)}{\psi(z)} < A$$

And we write

$$\phi = o(\psi) \qquad \text{as } z \to z_0$$

if

$$\lim_{z \to z_0} \frac{\phi(z)}{\psi(z)} = 0$$

---

[96] Usually in this case the terms of the series first decrease rapidly (the more rapidly the closer the independent variable is to its limiting value) but eventually start to increase again.

The symbols O and o are called *order symbols*, and in both cases we say that ϕ *is of order* ψ. Thus,

$$z^4 = O(z^5) \qquad \text{as } z \to \infty$$

$$z^4 = o(z^5) \qquad \text{as } z \to \infty$$

$$z^4 = O(4z^4 + 3z^3) \qquad \text{as } z \to \infty$$

but

$$z^4 \neq o(4z^4 + 3z^3) \qquad \text{as } z \to \infty$$

Also

$$z^5 = o(z^4) \qquad \text{as } z \to 0$$

but

$$z^4 \neq o(z^5) \qquad \text{as } z \to 0$$

and, for $\mathscr{R}e \, z > 0$,

$$e^{-z} = o(z^a) \qquad \text{as } \to \infty$$

for any constant a.

If $f(z)$ is an analytic function which has a pole of order m at z_0, then

$$f(z) = O((z - z_0)^{-m}) \qquad \text{and} \qquad f(z) \neq o((z - z_0)^{-m}) \qquad \text{as } z \to z_0$$

It is common practice to denote any sequence, say $a_1, a_2, a_3, \ldots$, by $\{a_j\}$. A sequence of functions $\{\phi_j(z)\}$ is said to be an *asymptotic sequence* as $z \to z_0$ if for each integer j, $\phi_{j+1} = o(\phi_j)$ as $z \to z_0$. Thus, $\{(z - z_0)^j\}$ is an asymptotic sequence as $z \to z_0$; $\{z^{-j}\}$ is as asymptotic sequence as $z \to \infty$; and $\{e^{-jz}\}$ is an asymptotic sequence as $z \to \infty$, provided $\mathscr{R}e \, z > 0$.

Note that equation (9–28) can be written in terms of the order symbol o as

$$w_1 - s_m = o(z^m) \qquad \text{as } z \to 0$$

We are, therefore, led to the following definition: Let $\{\phi_j(z)\}$ be an asymptotic sequence as $z \to z_0$. Then the formal series

$$\sum_{j=0}^{\infty} a_j \phi_j(z) \qquad\qquad (9\text{–}29)$$

with partial sums $S_m(z) = \sum_{j=0}^{m} a_j \phi_j(z)$ is called an *asymptotic series*. The asymptotic series (9–29) is said to be an *asymptotic expansion to m terms* of a function $f(z)$ if

$$f(z) - S_m(z) = o(\phi_m) \qquad \text{as } z \to z_0 \qquad\qquad (9\text{–}30)$$

and we write

$$f(z) \sim \sum_{j=0}^{m} a_j \phi_j(z)$$

If equation (9–30) holds for *every* positive integer m, we simply say that equation (9–29) is an *asymptotic expansion* of $f(z)$ and we write

$$f(z) \sim \sum_{j=0}^{\infty} a_j \phi_j(z)$$

Thus, in the preceding example, we have shown that

$$\int_0^{\infty} \frac{e^{-t} dt}{1 + zt} \sim \sum_{n=0}^{\infty} (-1)^n n! z^n$$

This series is typical in that most asymptotic series encountered in practice involve powers or inverse powers of z. Such series are called *asymptotic power series*.

Once a particular asymptotic sequence has been specified, any given function will have only one asymptotic expansion in terms of this sequence.

In this sense the asymptotic expansion is unique. However, two different functions may have the same asymptotic expansion. Thus, if

$$f(z) \sim \sum_{n=0}^{\infty} a_n z^{-n} \qquad \text{as } z \to \infty$$

it is not hard to verify that

$$f(z) + e^{-z} \sim \sum_{n=0}^{\infty} a_n z^{-n} \qquad \text{as } z \to \infty$$

for $\mathscr{Re}\ z > 0$.

It can be shown that if $f(z) \sim \sum a_n \phi_n(z)$ and $g(z) \sim \sum b_n \phi_n(z)$, then

$$\alpha f(z) + \beta g(z) \sim \sum (\alpha a_n + \beta b_n) \phi_n(z)$$

for any complex constants α and β. It is generally possible to integrate asymptotic expansions term by term, but differentiation is not always permissible (ref. 32).

Asymptotic power series can be manipulated in much the same way as convergent power series. Let the function $f(z)$ be a single-valued analytic function at every finite point z with $|z| > R$. If $f(z)$ has the asymptotic expansion

$$f(z) \sim \sum_{n=0}^{\infty} a_n z^{-n} \qquad \text{as } z \to \infty$$

and this expansion is valid for all values of the argument of z, then $f(z)$ must be bounded as $z \to \infty$. Hence, it cannot possess a pole or an essential singularity at $z = \infty$. Therefore, an analytic function cannot possess a single asymptotic expansion in inverse powers of z as $z \to \infty$ which is valid for all values of the argument of z unless it is analytic at infinity. Thus, asymptotic power series are frequently valid only over a given sector of the complex plane, and a different asymptotic representation must be used outside this sector. This is closely related to the so-called *Stokes phenomenon* which we shall encounter subsequently.

If a function $f(z)$ is analytic at a point z_0 and, therefore, possesses a convergent power series about z_0, it is not hard to see that the series must also be an asymptotic expansion of $f(z)$ as $z \to z_0$ (ref. 32, p. 22).

9.5 NORMAL SOLUTIONS

Let z_0 be an irregular singular point of the equation

$$w'' + p(z)w' + q(z)w = 0 \qquad (9\text{--}31)$$

and suppose that, if z_0 is a finite point,

$$p = O((z-z_0)^{-m}) \qquad \text{and} \qquad q = O((z-z_0)^{-(k+1)}) \qquad \text{as } z \to z_0$$

and that, if $z_0 = \infty$,

$$p = O(z^m) \qquad \text{and} \qquad q = O(z^{(k-1)}) \qquad \text{as } z \to \infty$$

where m and k are integers.

We have seen that, when $k \leq m$ (and only in this case), it is always possible to construct a formal series solution $w_1(z)$ of the form

$$w_1(z) = (z-z_0)^\rho \sum_{n=0}^{\infty} a_n(z-z_0)^n \quad \text{if } z_0 \text{ is finite} \qquad (9\text{--}32\text{a})$$

or of the form

$$w_1(z) = z^\rho \sum_{n=0}^{\infty} a_n z^{-n} \qquad \text{if } z_0 = \infty \qquad (9\text{--}32\text{b})$$

which satisfies the differential equation term by term. The coefficients of these expansions can be calculated recursively from the differential equation in much the same manner as those in the series solutions at regular singular points. These expansions will, in general, be divergent. However, as we have seen by example they actually turn out to be asymptotic expansions of a true solution to equation (9–31) at the point $z=z_0$. They will, therefore, be very useful for numerical computation [97] of this solution for z near z_0.

The series in equations (9–32a) and (9–32b) will not diverge only if they terminate after a finite number of terms, in which case these equations will represent regular solutions to the differential equation (9–31). A second solu-

---

[97] In fact, asymptotic expansions are quite frequently more suitable for numerical computation than convergent power series since fewer terms are required to obtain a given accuracy.

tion at the point z_0 can then always be found from equation (6–63). Thus, except in the relatively rare case where $k \leqslant m$ and the formal power series solution terminates, it is still necessary to find one or two solutions to equation (9–31) or at least the asymptotic expansions of these solutions about the irregular singular point $z = z_0$. To this end, we introduce into equation (9–31) the change of dependent variable

$$w = e^{\Omega} u \tag{9-33}$$

where

$$\Omega = \begin{cases} \sum_{n=1}^{s} \omega_n (z - z_0)^{-n} & \text{if } z_0 \text{ is finite} \\ \sum_{n=1}^{s} \omega_n z^n & \text{if } z_0 = \infty \end{cases} \tag{9-34}$$

and the coefficients ω_n for $n = 1, 2, \ldots, s$ are complex constants. Then equation (9–31) is transformed into the equation

$$u'' + p^+(z)u' + q^+(z)u = 0 \tag{9-35}$$

where

$$p^+(z) = p(z) + 2\Omega' \tag{9-36}$$

and

$$q^+(z) = q(z) + p(z)\Omega' + \Omega'' + \Omega'^2 \tag{9-37}$$

Now if, when z_0 is finite, the terms in $\Omega(z)$ can be chosen to cancel out enough of the negative powers of $z - z_0$ which arise from $p(z)$ and $q(z)$ so that $p^+(z)$ has, at most, a simple pole and $q^+(z)$ has, at most, a pole of order two at z_0, then z_0 will be a regular singular point of equation (9–35). Similar remarks, of course, apply when $z_0 = \infty$. In either case, then, when this can be done, a fundamental set of solutions can always be found by the methods of chapter 6; and equation (9–33) will then provide a fundamental set of solutions of equation (9–31). But this usually cannot be done. More frequently, it is only possible to adjust the coefficients in Ω so that the resulting equation (9–35) will possess a formal solution either of the form (9–32a) (when z_0 is finite) or of the form (9–32b) (when z_0 is infinite). Then equation (9–31) will have a formal solution either of the form

$$w(z) = e^{\Omega(z)}(z-z_0)^\rho \sum_{n=0}^{\infty} a_n(z-z_0)^n \qquad \text{if } z_0 \text{ is finite} \qquad (9\text{–}38a)$$

or of the form

$$w(z) = e^{\Omega(z)} z^\rho \sum_{n=0}^{\infty} a_n z^{-n} \qquad \text{if } z_0 = \infty \qquad (9\text{–}38b)$$

These formal solutions, called *normal solutions*, either will terminate or they will diverge in which case they will be asymptotic expansions.

Before discussing this in more detail, it is convenient to introduce the following definition: If the coefficients $p(z)$ and $q(z)$ of the differential equation

$$w'' + p(z)w' + q(z)w = 0 \qquad (9\text{–}39)$$

have, at most, poles at z_0, the equation is said to be of *finite rank at z_0*. If z_0 is a finite point and equation (9–39) is of finite rank at z_0, the smallest number r such that both

$$p = O((z-z_0)^{-(r+1)}) \qquad \text{and} \qquad q = O((z-z_0)^{-2(r+1)}) \qquad \text{as } z \to z_0$$

is called the rank of equation (9–39) at z_0. The rank of equation (9–39) at $z = \infty$ is defined to be the smallest number r such that

$$p = \bar{O}(z^{r-1}) \qquad \text{and} \qquad q = O(z^{2(r-1)}) \qquad \text{as } z \to \infty$$

provided the equation is of finite rank at ∞. If the equation has a regular singular point at z_0, it is said to be of rank zero.

For example, since the coefficients of the equation

$$zw'' + w' + w = 0 \qquad (9\text{--}40)$$

satisfy the conditions

$$\left.\begin{array}{ll} p = O(z^{-1}) & q = O(z^{-1}) \\[2mm] p \neq o(z^{-1}) & q \neq o(z^{-1}) \end{array}\right\} \text{ as } z \to \infty$$

we see that this equation is of rank 1/2 at $z = \infty$.

Now suppose that equation (9–31) is of rank $r > 0$ at the point z_0. In order that equation (9–35) possess a formal solution of the form (9–32a) when z_0 is a finite point, it is necessary that the highest negative power of $z - z_0$ in q^+ exceed the highest negative power in p^+ by no more than 1. And if equation (9–35) is to possess a formal solution of the form (9–32b) when $z_0 = \infty$, it is necessary that the highest power of z in q^+ be less than the highest power of z in p^+. Suppose that $z_0 = \infty$. An examination of the expression obtained by substituting equation (9–34) into equations (9–36) and (9–37) shows that q^+ will always contain higher powers of z than p^+ unless the numbers ω_n and s in equation (9–34) can be chosen to cancel out the coefficients of those powers of z in q^+ which are larger than, or equal to, those in p^+. This can be done only if the rank r of equation (9–31) is an integer. And the maximum value of s necessary to accomplish this will then always be less than, or equal to, r. The same conclusion applies when z_0 is finite. It is usually possible to choose the numbers ω_n and s so that the cancellation is accomplished in more than one way. In this case, we obtain two solutions to equation (9–31).

Thus, *equation (9–31) will possess formal solutions either of the form (9–38a) or the form (9–38b) at the point z_0 only if the rank r of this equation at z_0 is an integer. The function $\Omega(z)$ is then given by equation (9–34) with s ≤ r.*

Since equation (9–40) is of rank 1/2 at $z = \infty$, it does *not* possess a normal solution at this point.

As an example of an equation which does possess such a solution at $z = \infty$, consider the equation

$$w'' + q(z)w = 0 \qquad (9\text{--}41)$$

with

$$q(z) = \sum_{n=0}^{\infty} q_n z^{-n} \tag{9-42}$$

and

$$q_0 \neq 0 \tag{9-43}$$

We see in view of condition (9–43) that equation (9–41) is of rank 1 at $z=\infty$. Hence, we make the change of dependent variable

$$w = e^{\Omega} u \tag{9-44}$$

where, since $s \leq r = 1$,

$$\Omega = \omega_1 z \tag{9-45}$$

And upon substituting equations (9–44) and (9–45) into equation (9–41) we find that

$$u'' + p^+(z)u' + q^+(z)u = 0 \tag{9-46}$$

where $p^+(z) = 2\omega_1$ and $q^+(z) = \omega_1^2 + q_0 + \sum_{n=1}^{\infty} q_n z^{-n}$. The highest power of z which occurs in both p^+ and q^+ is zero. Hence, the highest power of z which occurs in q^+ can be made less than the highest power which occurs in p^+ by choosing ω_1 to be a root of the equation

$$\omega_1^2 + q_0 = 0$$

But the two roots of this equation are $\omega_{1,1} = i q_0^{1/2}$ and $\omega_{1,2} = -i q_0^{1/2}$. And each of these roots leads to a different form of equation (9–46). Thus, we obtain the two equations

$$u_1'' + 2iq_0^{1/2}u_1' + \tilde{q}(z)u_1 = 0 \qquad (9\text{-}46\text{a})$$

$$u_2'' - 2iq_0^{1/2}u_2' + \tilde{q}(z)u_2 = 0 \qquad (9\text{-}46\text{b})$$

where $\tilde{q}(z) = \sum_{n=1}^{\infty} q_n z^{-n}$. But we know that these equations possess formal solutions of the form

$$u_1 = z^{\rho_1} \sum_{n=0}^{\infty} a_n z^{-n} \qquad (9\text{-}47\text{a})$$

and

$$u_2 = z^{\rho_2} \sum_{n=0}^{\infty} b_n z^{-n} \qquad (9\text{-}47\text{b})$$

where ρ_1 and ρ_2 are each solutions of linear indicial equations and the coefficients a_n and b_n can be calculated successively from a recurrence relation of the usual type.

Thus, for example, upon substituting equation (9–47a) into equation (9–46a), shifting the indices, and collecting coefficients of like powers of z in the usual way, we obtain

$$\sum_{n=2}^{\infty} (\rho_1 - n + 2)(\rho_1 - n + 1)a_{n-2}z^{-n}$$

$$+ \sum_{n=1}^{\infty} z^{-n}\left[2iq_0^{1/2}(\rho_1 - n + 1)a_{n-1} + \sum_{k=1}^{n} q_k a_{n-k} \right] = 0$$

And after equating to zero the coefficients of like powers of z, we find that for $n = 1$ (since $a_0 \neq 0$)

$$2iq_0^{1/2}\rho_1 + q_1 = 0 \qquad (9\text{-}48)$$

and

$$[2iq_0^{1/2}(\rho_1-n+1)+q_1]a_{n-1}=-(\rho_1-n+2)\ (\rho_1-n+1)a_{n-2}$$

$$-\sum_{k=2}^{n}q_k a_{n-k}\qquad \text{for } n=2, 3, \ldots \qquad (9\text{--}49)$$

Equation (9–48) is the usual linear indicial equation. And substituting this into the recurrence relation (9–49) shows that

$$-8iq_0^{3/2}(n-1)a_{n-1}=[q_1+2iq_0^{1/2}(n-2)]\ [q_1+2iq_0^{1/2}(n-1)[a_{n-2}$$

$$-4q_0\sum_{k=2}^{n}q_k a_{n-k}\qquad \text{for } n=2, 3, \ldots \qquad (9\text{--}49\text{a})$$

This recurrence relation will either terminate or else it will determine the a_n so that the series (9–47a) diverges.

In any event, the two formal solutions (9–47a) and (9–47b) will now provide, through equations (9–44) and (9–45), two normal solutions to equation (9–41), say w_1 and w_2, which are given by

$$\left.\begin{aligned}w_1&=e^{iq_0^{1/2}z}\ z^{\rho_1}\sum_{n=0}^{\infty}a_n z^{-n}\\[2mm] w_2&=e^{-iq_0^{1/2}z}\ z^{\rho_2}\sum_{n=0}^{\infty}b_n z^{-n}\end{aligned}\right\}\qquad (9\text{--}50)$$

Note that if condition (9–43) did not hold (i.e., if $q_0=0$) but the condition $q_1\neq 0$ did hold, then equation (9–41) would be of rank 1/2 at $z=\infty$. In this case, it would not be possible to carry out the procedure just described.

9.6 SUBNORMAL SOLUTIONS

We have seen that if equation (9–31) is of rank r with $r > 0$ at the point z_0, it will have a formal solution of the form (9–38a) or (9–38b) only if r is an integer. However, a change of independent variable of the type

$$\zeta = \begin{cases} (z - z_0)^{1/k} & \text{if } z_0 \text{ is finite} \\ z^{1/k} & \text{if } z_0 = \infty \end{cases}$$

where k is a positive integer, will usually transform an equation of fractional rank into one ,whose rank is an integer. The formal solutions obtained by this procedure are called *subnormal* solutions.

Thus, we have shown that equation (9–41) in the preceding example has rank 1/2 at ∞ when condition (9–43) does not hold but the condition $q_1 \neq 0$ does hold. That is, the equation

$$w'' + q(z)w = 0 \tag{9–51}$$

where

$$q(z) = \sum_{n=1}^{\infty} q_n z^{-n} \tag{9–52}$$

and

$$q_1 \neq 0 \tag{9–53}$$

has rank 1/2 at $z = \infty$ and, therefore, does not possess a normal solution at this point. However, upon making the change of variable $\zeta = z^{1/2}$, equation (9–51) becomes

$$\frac{d^2 w}{d\zeta^2} - \frac{1}{\zeta}\frac{dw}{d\zeta} + Q(\zeta)w = 0 \tag{9–54}$$

where

$$Q(\zeta) = 4 \sum_{n=1}^{\infty} \zeta^{-2(n-1)} q_n \tag{9-55}$$

And since this equation is of rank 1 at $\zeta = \infty$, we introduce the change of variable

$$w = e^{\omega_1 \zeta} u \tag{9-56}$$

to obtain the equation

$$\frac{d^2 u}{d\zeta^2} + P_1(\zeta) \frac{du}{d\zeta} + Q_1(\zeta) u = 0 \tag{9-57}$$

where

$$P_1(\zeta) = -\frac{1}{\zeta} + 2\omega_1$$

$$Q_1(\zeta) = -\frac{\omega_1}{\zeta} + \omega_1^2 + 4q_1 + 4 \sum_{n=2}^{\infty} q_n \zeta^{-2(n-1)}$$

Then by putting

$$\omega_1^2 + 4q_1 = 0 \tag{9-58}$$

we find that the highest power of ζ which occurs in Q_1 is less than the highest power which occurs in P_1. And, as in the preceding example, corresponding to the two roots $\pm i 2 q_1^{1/2}$ of equation (9–58) we obtain the following two equations from equation (9–56):

$$\frac{d^2 u_1}{d\zeta^2} - \left(\frac{1}{\zeta} - 4i q_1^{1/2} \right) \frac{du_1}{d\zeta} + 4 \left[\frac{i q_1^{1/2}}{2\zeta} + \sum_{n=2}^{\infty} q_n \zeta^{-2(n-1)} \right] u_1 = 0$$

$$\frac{d^2u_2}{d\zeta^2} - \left(\frac{1}{\zeta} + 4iq_1^{1/2} \right) \frac{du_2}{d\zeta} + 4 \left[\frac{-iq_1^{1/2}}{2\zeta} + \sum_{n=2}^{\infty} q_n \zeta^{-2(n-1)} \right] u_2 = 0$$

But we know that these equations possess the formal solutions $u_1 = \zeta^{\rho_1} \sum_{n=0}^{\infty} a_n \zeta^{-n}$ and $u_2 = \zeta^{\rho_2} \sum_{n=0}^{\infty} b_n \zeta^{-n}$, whose coefficients can be calculated recursively in the usual way. And therefore equation (9–51) possesses the two subnormal solutions

$$w_1 = e^{i2q_1^{1/2}z^{1/2}} z^{(\rho_1/2)} \sum_{n=0}^{\infty} a_n z^{-n/2} \qquad (9\text{–}59)$$

$$w_2 = e^{-i2q_1^{1/2}z^{1/2}} z^{(\rho_2/2)} \sum_{n=0}^{\infty} b_n z^{-n/2} \qquad (9\text{–}60)$$

9.7 NATURE OF RIGOROUS PROOFS

The formal solutions obtained in the preceding section are asymptotic expansions of certain solutions to the differential equations. One method for actually proving this was developed by G. Birkhoff. In this method the leading terms of the partial sums of each formal solution are used to construct a homogeneous differential equation whose solution is known and which in a certain sense is close to the given equation when z is near [98] z_0. This equation is then used to construct a singular integral equation whose solution also satisfies the given differential equation. It is then proved that the solution to this integral equation possesses an asymptotic expansion which coincides with the formal solution obtained by the methods discussed previously. For a detailed discussion of this method, in which equation (9–41) is treated to illustrate the general principles, the reader is referred to reference 32.

9.8 CONNECTION OF SOLUTIONS: STOKES PHENOMENON

In the preceding examples, we have shown how to obtain the asymptotic expansions as $z \rightarrow \infty$ of two solutions to a differential equation of the type

---

[98] Without loss of generality, the point z_0 (about which the solutions are obtained) is taken as ∞.

(9–41). The problem frequently arises of connecting these asymptotic expansions with an expression [99] for some given solution to the differential equation which is valid for small values of z. We know, in general, that the asymptotic expansion of any given solution w to equation (9–41) can be expressed as a linear combination of the two asymptotic solutions (9–50). However, it usually turns out that the particular linear combination of these two asymptotic solutions used to represent the asymptotic expansion of the given solution w must be changed as the variable z crosses certain "critical rays" in going from one sector of the complex plane to another. This is the *Stokes phenomenon* which we have mentioned previously.

For example, the change of variable

$$W = z^{-1/2} w \tag{9–61}$$

transforms Bessel's equation

$$z^2 W'' + z W' + (z^2 - p^2) W = 0 \tag{9–62}$$

into the equation

$$w'' + \left(1 + \frac{\frac{1}{4} - p^2}{z^2} \right) w = 0 \tag{9–63}$$

which has the form of equation (9–41) with the coefficients in equation (9–42) given by

$$q_0 = 1 \qquad q_1 = 0 \qquad q_2 = \frac{1}{4} - p^2$$

and

$$q_n = 0 \quad \text{for } n = 3, 4, 5, \ldots \tag{9–64}$$

[99] Such as a power series expansion.

And if we choose the arbitrary constants a_0 and b_0, respectively, to be

$$\left.\begin{array}{l} a_0 = \left(\dfrac{2}{\pi}\right)^{1/2} \dfrac{1}{2}\, e^{-i[(p\pi/2)+(\pi/4)]} \\[3mm] b_0 = \left(\dfrac{2}{\pi}\right)^{1/2} \dfrac{1}{2}\, e^{i[(p\pi/2)+(\pi/4)]} \end{array}\right\} \tag{9-65}$$

it is easy to show that the formal solutions (9–50) become in this case

$$w_1 = \left(\frac{2}{\pi}\right)^{1/2} \frac{1}{2}\, e^{i[z-(p\pi/2)-(\pi/4)]} \sum_{n=0}^{\infty} a_n(-i)^n z^n \tag{9-66}$$

$$w_2 = \left(\frac{2}{\pi}\right)^{1/2} \frac{1}{2}\, e^{-i[z-(p\pi/2)-(\pi/4)]} \sum_{n=0}^{\infty} a_n(i)^n z^n \tag{9-67}$$

where

$$a_n = \frac{\left(\dfrac{1}{2}-p\right)_n \left(\dfrac{1}{2}+p\right)_n}{2^n n!}$$

Hence, the asymptotic expansion of any given solution W of equation (9–62) can be expressed as a linear combination

$$W(z) \sim C_1 z^{-1/2} w_1(z) + C_2 z^{-1/2} w_2(z)$$

of the functions (9–66) and (9–67) multiplied by $z^{-1/2}$. Let us choose the solution $W(z)$ to be the Bessel function of the first kind $J_p(z)$. Then equation (8–37) shows that for $-\pi < \arg z < \pi$, $C_1 = C_2 = 1$ and

$$J_p(z) \sim z^{-1/2} w_1(z) + z^{-1/2} w_2(z) \qquad -\pi < \arg z < \pi \tag{9-68}$$

This representation does not apply along the critical ray $\arg z = \pi$, that is, along the negative real axis. To find an asymptotic expansion which is valid in a region which includes the negative real axis, put

$$z = e^{\pi i} z_1 \qquad\qquad (9\text{--}69)$$

Then since arg $z = \pi + $ arg z_1, arg z varies between 0 and 2π as arg z_1 varies between $-\pi$ and π. And it follows from the definition (8–27) of the Bessel function that

$$J_p(z) = J_p(e^{\pi i} z_1) = e^{\pi p i} J_p(z_1) \qquad \text{for } |\arg z_1| < \pi$$

But arg z_1 is in the range where J_p has the asymptotic representation (9–68). Hence,

$$J_p(z) \sim e^{\pi i p} z_1^{-1/2} w_1(z_1) + e^{\pi i p} z_1^{-1/2} w_2(z_1) \qquad \text{for } |\arg z_1| < \pi$$

And since equations (9–66) and (9–67) show that

$$w_1(z_1) = e^{-i[p\pi + (\pi/2)]} w_2(z) \qquad\qquad w_2(z_1) = e^{i[p\pi + (\pi/2)]} w_1(z)$$

it follows that

$$J_p(z) \sim z^{-1/2} w_2(z) - e^{2\pi i p} z^{-1/2} w_1(z) \qquad \text{for } 0 < \arg z < 2\pi \qquad (9\text{--}70)$$

Comparing equations (9–68) and (9–70) shows that different linear combinations of $w_1(z)$ and $w_2(z)$ must be used to represent the asymptotic expansion of $J_p(z)$ in the sectors $-\pi < \arg z < \pi$ and $0 < \arg z < 2\pi$. And, equations (9–66), (9–67), and (9–70) show that

$$J_p(z) \sim e^{\pi i p + (3\pi i/2)} \left(\frac{2}{\pi z} \right)^{1/2} \sin\left(z + \frac{p\pi}{2} - \frac{\pi}{4} \right)$$

$$\text{to one term, for } 0 < \arg z < 2\pi \qquad (9\text{--}71)$$

Since in each region of the complex plane one of the functions, $w_1(z)$ and $w_2(z)$, will be exponentially small compared with the other, the expansions (9–69) and (9–70) will be equal to one another, with exponentially small error, in every region where they are both defined.

CHAPTER 10

Expansions in Small and Large Parameters: Singular Perturbations

Many of the differential equations encountered in practice contain parameters. Although it frequently happens that we cannot find the exact solutions to these equations, the behavior of these solutions for large or small values of the parameters is often physically important. In such cases we are content to find asymptotic representations of the solutions which are valid for these limiting values of the parameters. A number of techniques for obtaining such expansions are given in this chapter. Since no formal theory has been developed for many of these techniques, the approach of this chapter is necessarily heuristic and the material is frequently presented by means of examples. On the other hand, the ideas developed herein are quite general; and they apply not only to ordinary differential equations, but also to partial differential equations, integral equations, and even difference equations.

10.1 NONSINGULAR EXPANSIONS OF SOLUTIONS

The general second-order homogeneous linear equation containing a large parameter λ is of the form

$$\frac{d^2y}{dx^2} + p(x, \lambda)\, \frac{dy}{dx} + q(x, \lambda)y = 0 \tag{10-1}$$

We suppose that $p(x, \lambda)$ and $q(x, \lambda)$ have (at least formal) power series expansions in λ. If these expansions contain only nonpositive powers of λ, say

$$p(x, \lambda) = \sum_{n=0}^{\infty} p_n(x)\lambda^{-n} \tag{10-2}$$

$$q(x, \lambda) = \sum_{n=0}^{\infty} q_n(x)\lambda^{-n} \tag{10-3}$$

then the solutions to equation (10–1) will have a formal power series expansion of the form

$$y(x) = \sum_{n=0}^{\infty} y_n(x)\lambda^{-n} \tag{10-4}$$

where the functions $y_n(x)$ are determined by substituting the expansion (10–4) into equation (10–1) and equating to zero the coefficients of like powers of λ to obtain

$$\frac{d^2y_n}{dx^2} + p_0(x)\frac{dy_n}{dx} + q_0(x)y_n = -\sum_{k=0}^{n-1}\left[p_{n-k}(x)\frac{dy_k}{dx} + q_{n-k}(x)y_k \right]$$

$$\text{for } n=0, 1, 2, \ldots \tag{10-5}$$

and the sum on the right is omitted for $n=0$. These equations can, at least in principle, be solved successively for the coefficients y_n. If the series (10–2) and (10–3) converge, it can be shown (ref. 4, p. 126, ex. 6) that the series (10–4) will also converge. In any case, the formal series (10–4) will usually be an asymptotic expansion.

However, if the formal power series expansion for either $p(x, \lambda)$ or $q(x, \lambda)$ involves any positive powers of λ, the expansion of the solution in powers of λ will, in general, involve infinitely many positive and negative powers of λ; and it will not be possible to solve for the coefficients successively. Nevertheless, it is still possible in certain instances to obtain formal solutions whose terms can be calculated successively by using a technique analogous to the procedure used in the preceding chapter to obtain solutions at irregular singular points. These solutions will not, in general, be convergent series; but they will be asymptotic expansions. Thus, consider the equation

$$\frac{d^2y}{dx^2} + p(x, \lambda)\frac{dy}{dx} + q(x, \lambda)y = 0 \tag{10-6}$$

where

$$p(x, \lambda) = \sum_{n=0}^{\infty} p_n(x) \lambda^{k-n} \tag{10-7}$$

$$q(x, \lambda) = \sum_{n=0}^{\infty} q_n(x) \lambda^{2k-n} \tag{10-8}$$

k is a positive integer, and $p_0(x)$ and $q_0(x)$ are not both identically zero. Then it can be shown by direct substitution that equation (10–6) will possess two formal solutions of the form

$$y = e^{\Omega(x)} \sum_{n=0}^{\infty} y_n(x) \lambda^{-n} \tag{10-9}$$

where

$$\Omega(x) = \sum_{n=0}^{k-1} \omega_n(x) \lambda^{k-n} \tag{10-10}$$

The coefficents y_n and ω_n can be calculated by solving successively the set of ordinary differential equations which is obtained by substituting the assumed solution (10–9) into equation (10–6) and equating to zero the coefficients of like powers of λ.

In order to illustrate these ideas, consider the differential equation

$$\frac{d^2y}{dx^2} + \left[\lambda^2 q_0(x) + q_2(x) \right] y = 0 \tag{10-11}$$

which was first discussed by Liouville in his classical investigations of the Sturm-Liouville problems.

Since in this case $k = 1$, the differential equation will possess formal solutions of the form

$$y = e^{\omega_0(x)\lambda} \sum_{n=0}^{\infty} y_n(x) \lambda^{-n} \tag{10-12}$$

After substituting this into equation (10–11) and collecting terms, we find that

$$\sum_{n=0}^{\infty} (y_n'' + q_2 y_n)\lambda^{-n} + \sum_{n=0}^{\infty} (2\omega_0' y_n' + \omega_0'' y_n)\lambda^{1-n} + \sum_{n=0}^{\infty} (q_0 + \omega_0'^2)y_n\lambda^{2-n} = 0$$

But upon shifting indices, this becomes

$$\sum_{n=2}^{\infty} (y_{n-2}'' + q_2 y_{n-2})\lambda^{2-n} + \sum_{n=1}^{\infty} (2\omega_0' y_{n-1}' + \omega_0'' y_{n-1})\,\lambda^{2-n}$$

$$+ \sum_{n=0}^{\infty} (q_0 + \omega_0'^2)y_n\lambda^{2-n} = 0$$

And by equating the coefficients of the like powers of λ to zero, we get

$$(q_0 + \omega_0'^2)y_0 = 0 \qquad \text{for } n = 0 \tag{10–13}$$

$$2\omega_0' y_0' + \omega_0'' y_0 + (q_0 + \omega_0'^2)y_1 = 0 \qquad \text{for } n = 1 \tag{10–14}$$

and

$$y_{n-2}'' + q_2 y_{n-2} + 2\omega_0' y_{n-1}' + \omega_0'' y_{n-1} + (q_0 + \omega_0'^2)y_n = 0 \qquad \text{for } n = 2, 3, \ldots$$

$$\tag{10–15}$$

In order to avoid the trivial solution $y = 0$, we must require that $y_0 \neq 0$. Hence, equation (10–13) becomes

$$q_0 + \omega_0'^2 = 0 \tag{10–16}$$

When this is used in equations (10–14) and (10–15), we obtain

$$\left. \begin{aligned} 2\omega_0' y_0' + \omega_0'' y_0 &= 0 \\[2mm] 2\omega_0' y_{n-1}' + \omega_0'' y_{n-1} &= -y_{n-2}'' - q_2 y_{n-2} \qquad \text{for } n = 2, 3, \ldots \end{aligned} \right\} \tag{10–17}$$

In the case of the differential equation (10–1) we had to solve the set of second-order differential equations (10–5) to determine the coefficients y_n; whereas, in this case it is only necessary to solve a set of first-order equations.[100] Since $q_0 \neq 0$ equation (10–16) will have two distinct solutions, say $\omega_{0,1}$ and $\omega_{0,2}$, given by

$$\omega_{0,1} = + \int \sqrt{-q_0}\, dx \qquad (10\text{–}18)$$

$$\omega_{0,2} = - \int \sqrt{-q_0}\, dx \qquad (10\text{–}19)$$

Corresponding to each of these roots, we will obtain a different set of equations from equations (10–17) and, therefore, equation (10–12) will yield two different formal solutions to equation (10–11). It is proved in reference 32 (p. 83), that these two formal solutions are actually asymptotic expansions of two linearly independent solutions to equation (10–11) in any finite interval $a < x < b$ in which q_0 does not take on the value zero. The first equation (10–17) will yield the same equation for both of the roots (10–18) and (10–19). Thus, when equation (10–16) is substituted into the first equation (10–17), we obtain the separable equation

$$y_0' + \left(\frac{q_0'}{4q_0}\right) y_0 = 0$$

which is easily solved to obtain

$$y_0 = \text{constant} \times q_0^{-1/4} \qquad (10\text{–}20)$$

Equations (10–18) to (10–20) can now be substituted into equation (10–12) to show that the general solution of equation (10–11) in any interval $a < x < b$ where $q_0(x) \neq 0$ has the asymptotic expansion

$$y \sim C_1 q_0^{-1/4} e^{\lambda \int \sqrt{-q_0}\, dx} + C_2 q_0^{-1/4} e^{-\lambda \int \sqrt{-q_0}\, dx} \qquad \text{to one term} \qquad (10\text{–}21)$$

---

[100] This situation is analogous to the reduction of the indicial equation from a quadratic to a linear equation in going from a regular singular point to an irregular singular point.

where C_1 and C_2 are arbitrary constants. Thus, $q_0(x)$ must be either strictly positive or else strictly negative for all $a < x < b$. If q_0 is strictly negative, the constants of equation (10–21) can be redefined slightly to obtain

$$y \sim C_3(-q_0)^{-1/4}e^{\lambda \int \sqrt{-q_0}\,dx} + C_4(-q_0)^{-1/4}e^{-\lambda \int \sqrt{-q_0}\,dx}$$

$$\text{to one term, for } q_0(x) < 0 \qquad (10\text{–}22)$$

And if $q_0(x)$ is strictly positive for all $a < x < b$, the constants can be redefined to obtain

$$y \sim C_1 q_0^{-1/4} \sin\left(\lambda \int \sqrt{q_0}\,dx\right) + C_2 q_0^{-1/4} \cos\left(\lambda \int \sqrt{q_0}\,dx\right)$$

$$\text{to one term, for } q_0(x) > 0 \qquad (10\text{–}23)$$

Notice that the asymptotic solution (10–22) has a monotonic behavior, while the asymptotic solution (10–23) has an oscillatory behavior.

10.2 TRANSITION POINTS

Now let us consider the case where $q_0(x)$ is equal to zero at a single point, say x_0, in the interval $a < x < b$. And suppose, in addition, that $q_0(x)$ and $q_2(x)$ are analytic at x_0. Then $q_0(x)$ and $q_2(x)$ must have the expansions

$$q_0(x) = q_{0,1}(x - x_0) + q_{0,2}(x - x_0)^2 + \dots \qquad (10\text{–}24)$$

$$q_2(x) = q_{2,0} + q_{2,1}(x - x_0) + \dots \qquad (10\text{–}25)$$

about the point $x = x_0$, and we shall require for definiteness that $q_{0,1} \neq 0$.

Thus, the point x_0 is an ordinary point of the differential equation (10–11). Every solution to this equation must, therefore, be analytic at x_0. But in view of equation (10–24), equation (10–21) implies that

$$y = O((x - x_0)^{-1/4}) \qquad \text{as } x \to x_0 \qquad (10\text{–}26)$$

Hence, the asymptotic expansion (10–21) cannot represent an analytic function in a neighborhood of x_0 and, therefore, it certainly cannot be the asymptotic expansion of any solution to equation (10–11) in the neighborhood of this point.

However, for any positive number δ (no matter how small), equation (10–21) provides a valid asymptotic expansion, to one term, of the general solution to equation (10–11) both in the interval $a < x < x_0 - \delta$ and in the interval $x_0 + \delta < x < b$ because $q_0(x)$ does not vanish in either of these intervals. Since equation (10–24) shows that $q_0(x)$ is not tangent to the x-axis at x_0, it must cross the axis at this point and must, therefore, be positive in one of the intervals. The solution to equation (10–11) will then have the asymptotic expansion (10–23) in this interval. In the other interval, $q_0(x)$ will be negative and equation (10–11) will have the asymptotic solution (10–22). Thus, the asymptotic solution to equation (10–11) will have an oscillatory behavior on one side of x_0 and will have a monotonic behavior on the other side. The transition from one type of behavior to the other takes place in a small region centered at x_0. The point x_0 is, therefore, called a *transition point*.[101]

When the second equation (10–17) with $n=2$ is solved for $y_1(x)$ and equations (10–16) and (10–20) are substituted into the result, it is found from equations (10–24) and (10–25) that

$$y_1 = O\left((x-x_0)^{-7/4}\right) \qquad \text{as } x \to x_0$$

But equation (10–26) shows that y_0 is of order $(x-x_0)^{-1/4}$ as $x \to x_0$. Hence, the second term in the expansion (10–12) will be approximately equal to the first term when

$$\left|(x - x_0)^{-1/4}\right| \approx \lambda^{-1}\left|(x - x_0)^{-7/4}\right|$$

that is, when the distance between x and x_0 is approximately $|x - x_0| \approx \lambda^{-2/3}$.

Thus, when the distance between x and x_0 is less than or equal to $\lambda^{-2/3}$, the order of the terms of the expansion (10–12) will actually increase with increasing n. This shows that the asymptotic expansion breaks down in a region of radius $\lambda^{-2/3}$ about the point x_0. It is, therefore, said to be a *nonuniformly valid* asymptotic expansion.

In order to obtain an asymptotic expansion which is valid in this region, we rescale the independent variable so that it will be of order 1 therein. That is, we introduce the new independent variable

$$\xi = (x-x_0)\lambda^{2/3} \qquad\qquad (10\text{--}27)$$

---

[101] The term *turning point* is also used. This term arises from the application of eq. (10–11) to classical and quantum mechanical wave reflection problems.

into equations (10–11), (10–24), and (10–25) to get

$$\frac{d^2y}{d\xi^2}+\left[\lambda^{2/3}q_0\left(\frac{\xi}{\lambda^{2/3}}\right)+\frac{1}{\lambda^{4/3}}q_2\left(\frac{\xi}{\lambda^{2/3}}\right)\right]y=0 \qquad (10-28)$$

$$q_0\left(\frac{\xi}{\lambda^{2/3}}\right)=q_{0,1}\frac{\xi}{\lambda^{2/3}}+q_{0,2}\frac{\xi^2}{\lambda^{2/3}}+\ \cdot\ \cdot\ \cdot \qquad (10-29)$$

and

$$q_2\left(\frac{\xi}{\lambda^{2/3}}\right)=q_{2,0}+q_{2,1}\frac{\xi}{\lambda^{2/3}}+\ \cdot\ \cdot\ \cdot \qquad (10-30)$$

Then upon substituting equations (10–29) and (10–30) into equation (10–28) and neglecting terms which are small for large values of λ, we get

$$\frac{d^2y}{d\xi^2}+q_{0,1}\xi y=0 \qquad (10-31)$$

The solution to this equation should be "close" to the true solution to equation (10–11) at least in the region

$$|x-x_0|<\lambda^{-2/3} \qquad (10-32)$$

or $|\xi|<1$. Hence, it should represent the first term of the asymptotic expansion of the solution to equation (10–11) in the region (10–32).

The differential equation (10–31) is a form of Airy's equation (ref. 29, section 6.4) which can be transformed by the change of variable

$$y=\sqrt{\xi}\,Y \qquad\qquad \zeta=\frac{2}{3}\sqrt{q_{0,1}}\,\xi^{3/2}$$

into the Bessel's equation

$$\zeta^2 \frac{d^2Y}{d\zeta^2} + \zeta \frac{dY}{d\zeta} + \left[\zeta^2 - \left(\frac{1}{3} \right)^2 \right] Y = 0$$

(see eq. (8–24)). Thus, the general solution to equation (10–31) is

$$y = \left(\frac{\pi}{3} \lambda^{1/3} \right)^{1/2} D_1 \xi^{1/2} j_{1/3} \left(\frac{2}{3} \sqrt{q_{0,1}} \, \xi^{3/2} \right)$$

$$+ \left(\frac{\pi}{3} \lambda^{1/3} \right)^{1/2} D_2 \xi^{1/2} J_{-1/3} \left(\frac{2}{3} \sqrt{q_{0,1}} \xi^{3/2} \right) \qquad (10\text{–}33)$$

where D_1 and D_2 are arbitrary constants and the normalization factor $(\pi\lambda^{1/3}/3)^{1/2}$ has been inserted for convenience.

Let us now suppose, for definiteness, that

$$q_{0,\,1} > 0 \qquad\qquad (10\text{–}34)$$

Then solution (10–22) will hold in the region $a < x < x_0 - \delta$ and the solution (10–23) will hold in the region $x_0 + \delta < x < b$, where δ is some arbitrarily small positive constant. The solution (10–33) holds in some region which is centered at x_0 and has a size at least of order $\lambda^{-2/3}$. Having found the pieces of the solution which apply in the three regions into which the interval $a < x < b$ has been divided, it is now necessary to match up these pieces across the adjacent regions to form one continuous solution. These pieces of the solution contain altogether six arbitrary constants. Four of these will be determined by the matching requirement. The other two must remain arbitrary if we are to obtain an asymptotic expansion of the general solution to equation (10–11). In order to accomplish this matching, we suppose that the regions of validity of the various pieces of the solution can be extended in such a way that any two adjacent regions overlap one another. Thus, we assume there is an "*overlap domain*" (or *intermediate region*) in which both the expansions (10–22) and (10–33) are asymptotic expansions of the solution to equation (10–11) and that there is an overlap domain in which both the expansions (10–23) and (10–33) are asymptotic expansions of the solutions to equation (10–11) (see

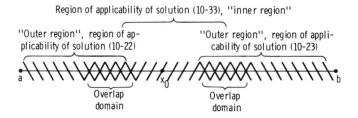

FIGURE 10–1. — Regions for transition-point expansion.

fig. 10–1). We therefore require that within the overlap domains the two adjacent expansions agree with each other to within an error which is of smaller order than the last term retained in these expansions.

Since the size of the region in which the solution (10–33) applies (which we will call the *inner region*) approaches zero as $\lambda \to \infty$, the location of the two overlap regions must also approach x_0 as $\lambda \to \infty$. Now in the outer regions, the asymptotic expansions correspond to holding the variable x fixed and taking the limit $\lambda \to \infty$. In the inner region, the asymptotic expansions correspond to holding the variable $\xi = (x - x_0)\lambda^{2/3}$ fixed and taking the limit $\lambda \to \infty$. This limiting process allows the coordinate x in the inner region to move toward x_0 as the size of this region shrinks to zero. Thus, the matching can be performed in a precise manner by introducing into both the inner and outer expansions an *"intermediate variable"*

$$\eta = (x - x_0)\lambda^{\alpha} = \xi\lambda^{[\alpha - (2/3)]} \tag{10–35}$$

with α chosen so that a fixed value of η will remain in the overlap region as this region shrinks toward x_0 with increasing λ. The matching of the adjacent expansions is then accomplished by changing their independent variables from x and ξ to η and then requiring that these expansions become identical (to the appropriate order in λ^{-1}) when the limit $\lambda \to \infty$ is taken while holding η fixed. However, if η is to lie in the overlap regions, we must require that

$$0 < \alpha < \frac{2}{3} \tag{10–36}$$

Now in order to apply the matching procedure to this problem, we introduce the new variable (10–35) into equation (10–33) to obtain

$$y = \left(\frac{\pi}{3}\lambda^{1-\alpha}\right)^{1/2} D_1 \eta^{1/2} J_{1/3}\left\{\frac{2}{3}\sqrt{q_{0,1}}\,\eta^{3/2}\lambda^{[(2/3)-\alpha]3/2}\right\}$$

$$+ \left(\frac{\pi}{3}\lambda^{1-\alpha}\right)^{1/2} D_2 \eta^{1/2} J_{-1/3}\left\{\frac{2}{3}\sqrt{q_{0,1}}\,\eta^{3/2}\lambda^{[(2/3)-\alpha]3/2}\right\} \quad (10\text{–}37)$$

First, suppose that $\eta > 0$. Then equation (8–37) shows that in the limit $\lambda \to \infty$ with η fixed

$$y \sim \left(\frac{\lambda^\alpha}{q_{0,1}\eta}\right)^{1/4} D_1 \cos\left\{\frac{2}{3}\sqrt{q_{0,1}}\,\eta^{3/2}\lambda^{[(2/3)-\alpha]3/2} - \frac{5\pi}{12}\right\}$$

$$+ \left(\frac{\lambda^\alpha}{q_{0,1}\eta}\right)^{1/4} D_2 \cos\left\{\frac{2}{3}\sqrt{q_{0,1}}\eta^{3/2}\lambda^{[(2/3)-\alpha]3/2} - \frac{\pi}{12}\right\} \quad \text{to one term}$$

$$(10\text{–}38)$$

But for $\eta < 0$ we find that $\eta^{3/2} = e^{i(3\pi/2)}|\eta|^{3/2}$; and therefore the argument of the Bessel's function in equation (10–37) is no longer in the range where equation (8–37) holds. We must, therefore, use equation (9–71).

In this case, we see that in the limit as $\lambda \to \infty$ with η fixed

$$y \sim \frac{1}{2}\left(\frac{\lambda^\alpha}{q_{0,1}|\eta|}\right)^{1/4}(D_2 - D_1)\exp\left\{\frac{2}{3}\sqrt{q_{0,1}}\,|\eta|^{3/2}\lambda^{[2/3-\alpha]3/2}\right\}$$

$$+ \frac{1}{2}\left(\frac{\lambda^\alpha}{q_{0,1}|\eta|}\right)^{1/4}(D_1 e^{-i\pi/6} + D_2 e^{i\pi/6})\exp\left\{-\frac{2}{3}\sqrt{q_{0,1}}\,|\eta|^{3/2}\lambda^{[2/3-\alpha]3/2}\right\}$$

$$\text{to one term} \quad (10\text{–}39)$$

Next substituting equation (10–35) into equation (10–23) and neglecting the higher order terms in λ^{-1}, we find that

$$y \sim \left(\frac{\lambda^\alpha}{q_{0,1}\eta} \right)^{1/4} C_1 \sin \left\{ \frac{2}{3} \sqrt{q_{0,1}} \, \eta^{3/2} \lambda^{[(2/3)-\alpha]3/2} \right\}$$

$$+ \left(\frac{\lambda^\alpha}{q_{0,1}\eta} \right)^{1/4} C_2 \cos \left\{ \frac{2}{3} \sqrt{q_{0,1}} \, \eta^{3/2} \lambda^{[(2/3)-\alpha]3/2} \right\} \qquad (10\text{–}40)$$

Hence, the expansion (10–38) and the expansion (10–40) will become identical in the right overlap domain if

$$C_1 = D_1 \sin \frac{5\pi}{12} + D_2 \sin \frac{\pi}{12}$$

$$C_2 = D_1 \cos \frac{5\pi}{12} + D_2 \cos \frac{\pi}{12}$$

and the expansion (10–23) becomes

$$y \sim D_1 q_0^{-1/4} \cos \left(\lambda \int_{x_0}^x \sqrt{q_0} \, dx - \frac{5\pi}{12} \right) + D_2 q_0^{-1/4} \cos \left(\lambda \int_{x_0}^x \sqrt{q_0} \, dx - \frac{\pi}{12} \right)$$

$$\text{to one term, for } x_0 < x < b \qquad (10\text{–}41)$$

In order to match the solutions in the left overlap domain, we must consider two cases. First, suppose that $D_1 \neq D_2$. Then the second term in equation (10–39) is negligibly small compared with the first and may be neglected. Hence, equation (10–39) becomes

$$y \sim \frac{1}{2} \left(\frac{\lambda^\alpha}{q_{0,1}|\eta|} \right)^{1/4} (D_2 - D_1) \exp \left\{ \frac{2}{3} \sqrt{q_{0,1}} \, |\eta|^{3/2} \lambda^{[(2/3)-\alpha]3/2} \right\}$$

$$(10\text{–}42)$$

This must now match with equation (10–22) in the left overlap domain. It can be seen that this matching can occur only if $C_3 \neq 0$. In this case, the second term will be exponentially small, compared with the first; and equation

292

(10–22) becomes, upon substituting in equation (10–35) and neglecting higher order terms in $1/\lambda$ while holding η fixed,

$$y \sim C_3 \left(\frac{\lambda^\alpha}{q_{0,1}|\eta|} \right)^{1/4} \exp \left\{ \frac{2}{3} \sqrt{q_{0,1}} \, |\eta|^{3/2} \lambda^{[(2/3)-\alpha]3/2} \right\}$$

But this will match with equation (10–42) in the left overlap domain only if $C_3 = (D_2 - D_1)/2$. And the solution (10–22) therefore becomes

$$y \sim \frac{D_2 - D_1}{2} (-q_0)^{-1/4} \exp \left(\lambda \int_{x_0}^{x} \sqrt{-q_0} \, dx \right) \qquad \text{to one term,}$$

$$\text{for } a < x < x_0 \text{ and } D_1 \neq D_2 \qquad (10\text{–}43)$$

Next consider the case where $D_1 = D_2$. Then the first term in equation (10–39) is zero and, therefore, the second term cannot be neglected. Equation (10–22) will only match onto equation (10–39) in the left overlap domain if $C_3 = 0$ and $C_4 = (\sqrt{3}/2) D_1$. Therefore, equation (10–22) becomes

$$y \sim \frac{\sqrt{3}}{2} D_1 (-q_0)^{-1/4} \exp \left(-\lambda \int_{x_0}^{x} \sqrt{-q_0} \, dx \right) \qquad \text{to one term,}$$

$$\text{for } a < x < x_0 \text{ and } D_1 = D_2 \qquad (10\text{–}44)$$

Thus, the complete asymptotic expansion of the general solution to equation (10–11) in the interval $a < x < b$ is given by equations (10–33), (10–41), and (10–43) or (10–44). The expansions (10–41) and (10–43) or (10–44) become poorer and poorer representations of the true solution as $x \to x_0$ (they are nonuniformly valid expansions), but the expansion (10–33) provides a good representation in this region. It is possible to obtain a uniformly valid asymptotic expansion in the region $x_0 \leqslant x < b$ by adding the expansion (10–41) to the expansion (10–33) and (so that it will not be counted twice) subtracting their common value in the transition region given by equation (10–38). Thus,[102]

$$y_{\text{uniformly valid}} \sim \text{eq. } (10\text{–}41) + \text{eq. } (10\text{–}33) - \text{eq. } (10\text{–}38)$$

---

[102] By construction, eqs. (10–33) and (10–38) are asymptotically equal in the outer region.

Similar remarks, of course, apply for the interval $a < x \leqslant x_0$. Although this method is very general and can be used even when a larger number of terms are retained in the asymptotic expansion, there is a better method of obtaining a uniformly valid one-term asymptotic expansion for the present problem which, in fact, applies to the entire interval $a < x < b$. To this end, notice that the term $\lambda \int_{x_0}^{x} \sqrt{q_0}\,dx$ becomes asymptotically equal to the term

$$\frac{2}{3}\lambda \sqrt{q_{0,1}}\,(x - x_0)^{3/2} = \frac{2}{3}\sqrt{q_{0,1}}\,\xi^{3/2}$$

in the transition region. It is, therefore, reasonable to hope that the range of validity of equation (10–33) (which applies for small values of [103] $x - x_0$) can be extended by replacing $2/3 \sqrt{q_{0,1}}\,\xi^{3/2}$ in the arguments of the Bessel functions by $\lambda \int_{x_0}^{x} \sqrt{q_0}\,dx$ and replacing the factor $\lambda^{1/6}\sqrt{\xi}$ by $q_0^{-1/4}\left(\dfrac{3\lambda}{2}\displaystyle\int_{x_0}^{x}\sqrt{q_0}\,dx\right)^{1/2}$. Upon making these substitutions, equation (10–33) becomes

$$y \sim q_0^{-1/4}\left(\frac{\pi}{2}\lambda \int_{x_0}^{x}\sqrt{q_0}\,dx\right)^{1/2}\left[D_1 J_{1/3}\left(\lambda \int_{x_0}^{x}\sqrt{q_0}\,dx\right) + D_2 J_{-1/3}\left(\lambda \int_{x_0}^{x}\sqrt{q_0}\,dx\right)\right]$$

$$(10\text{–}45)$$

Having obtained this result by a heuristic argument, it is now easy to verify that, for $\lambda^{2/3}(x - x_0) = O(1)$, this equation agrees with equation (10–33) to the lowest order λ^{-1} and, for $x - x_0 = O(1)$ with $x > x_0$, it agrees to lowest order in λ^{-1} with equation (10–41). Similarly, when $x - x_0 = O(1)$ and $x < x_0$, it agrees to lowest order in λ^{-1} with either equation (10–43) or equation (10–45), depending on whether D_1 is equal to D_2. Thus, equation (10–45) must represent a uniformly valid asymptotic expansion to one term of the general solution to equation (10–11). This result was first obtained by Langer (refs. 33 to 35) by a more formal procedure. The technique of using inner and outer expansions to obtain a uniformly valid expansion began with Friedrichs (ref. 36) in the 1950's and was developed by Kaplan, Lagerstrom, Cole, and many others. This technique not only applies to linear and nonlinear ordinary differential equations but also to partial differential equations. Many of the applications of this method have been to fluid mechanics problems. In fact, the basic ideas grew out of boundary layer theory.

---

[103] This is because ξ is of order 1 and, therefore, for large λ, x must be close to x_0.

10.3 MATCHED ASYMPTOTIC EXPANSIONS

The method of using inner and outer expansions to obtain a uniformly valid expansion can, in fact, be applied to obtain solutions in a systematic way to certain equations which are of the form (10–6). The solution (10–9) to equation (10–6) was obtained by essentially guessing its general form. Therefore, this procedure is limited to equations of the type (10−6); the method of matched asymptotic expansions has no such limitation.

The method is usually applied in such a way that it is necessary to consider the boundary conditions along with the differential equation. The ideas involved are best illustrated by means of an example. However, instead of considering an equation containing a large parameter we will now consider an equation containing a small parameter [104] ϵ. Thus, we shall seek an asymptotic expansion as $\epsilon \to 0$ of the solution to the equation

$$\epsilon \frac{d^2 y}{dx^2} + p(x) \frac{dy}{dx} + q(x) y = 0 \tag{10–46}$$

subject to the boundary conditions

$$y(0) = 1 \tag{10–47}$$

$$y(1) = a \tag{10–48}$$

and where the functions $p(x)$ and $q(x)$ are any functions which can be represented by power series, say

$$p(x) = \sum_{n=0}^{\infty} p_n x^n \qquad q(x) = \sum_{n=0}^{\infty} q_n x^n \tag{10–49}$$

near $x = 0$, and by similar power series, say

$$p(x) = \sum_{n=0}^{\infty} \tilde{p}_n (1-x)^n \qquad q(x) = \sum_{n=0}^{\infty} \tilde{q}_n (1-x)^n$$

---

[104] Of course, since we can always put $\lambda = 1/\epsilon$ to obtain an equation containing a large parameter, there is no real difference involved.

near $x=1$. We shall also require that

$$p(x) > 0 \qquad \text{for } 0 \leqslant x \leqslant 1 \tag{10–50}$$

and that the integral $\int_0^1 [q(x)/p(x)]dx$ exists.

Since ϵ is small, it is natural to seek a solution to equation (10–46) in the form of a power series in ϵ. Thus, let us try to obtain a formal solution to equation (10–46) of the form

$$y(x; \epsilon) = \sum_{n=0}^{\infty} y_n(x)\epsilon^n \tag{10–51}$$

Upon substituting this into equations (10–46) to (10–48) and equating to zero the coefficients of like powers of ϵ, we find that

$$p(x)\frac{dy_0}{dx} + q(x)y_0 = 0 \tag{10–52}$$

$$p(x)\frac{dy_n}{dx} + q(x)y_n = -\frac{d^2y_{n-1}}{dx^2} \qquad \text{for } n = 1, 2, \ldots \tag{10–53}$$

$$y_0(0) = 1 \tag{10–54}$$

$$y_0(1) = a \tag{10–55}$$

$$y_n(0) = y_n(1) = 0 \qquad \text{for } n = 1, 2, \ldots \tag{10–56}$$

Since equations (10–52) and (10–53) are first-order linear equations, they can be solved immediately to obtain

$$y_0(x) = C_0 e^{-\Omega(x)} \tag{10–57}$$

$$y_n(x) = e^{-\Omega(x)}\left[C_n + \int_1^x e^{\Omega(x)}\frac{1}{p(x)}\frac{d^2y_{n-1}}{dx^2}dx \right] \qquad \text{for } n = 1, 2, \ldots \tag{10–58}$$

where the C_n for $n = 0, 1, 2, \ldots$ are arbitrary constants and we have put

$$\Omega(x) \equiv \int_1^x \frac{q(x)}{p(x)}\, dx \tag{10–59}$$

Notice that since the integral $\Omega(0)$ exists by hypothesis, the integral (10–59) must certainly be finite for all x in the interval $0 \leqslant x \leqslant 1$.

Now, except in the very exceptional circumstance where $a = e^{\Omega(0)}$, it is impossible to choose C_0 so that the zeroth-order solution (10–57) satisfies both boundary conditions. Thus, suppose that $a \neq e^{\Omega(0)}$ and, therefore, that one of the boundary conditions (10–54) or (10–55) cannot be satisfied. We reason that, just as the expansion (10–12) broke down at the transition point, the expansion (10–51) will break down at one of the boundary points $x = 0$ or $x = 1$ and it will be necessary to obtain a different (inner) expansion in this region.[105] This expansion must then be matched smoothly onto the outer expansion (10–51) in some intermediate region. It is necessary to investigate the behavior of the asymptotic solutions at both boundaries in order to find which one will correspond to the boundary layer region. If this is done, it will be found that, due to the condition (10–50), it will be impossible to match any inner solution which occurs near the boundary $x = 1$ to the solutions (10–57) and (10–58). This is due to the fact that condition (10–50) will cause all possible solutions for the region near $x = 1$ to grow exponentially with the distance $1 - x$ and this type of behavior is incompatible with the solutions (10–57) and (10–58).

The constants C_n for $n = 0, 1, 2, \ldots$ must, therefore, be determined so that the boundary condition (10–55) and the second boundary condition (10–56) at $x = 1$ are satisfied. Hence, $C_0 = a$ and $C_n = 0$ for $n = 1, 2, \ldots$; and the solutions (10–57) and (10–58) become, respectively,

$$y_0(x) = a e^{-\Omega(x)} \tag{10–60}$$

$$y_n(x) = -e^{-\Omega(x)} \int_1^x e^{\Omega(x)} \frac{1}{p(x)} \frac{d^2 y_{n-1}}{dx^2}\, dx \quad \text{for } n = 1, 2, \ldots \tag{10–61}$$

The functions $y_n(x)$ for $n = 1, 2, \ldots$ can now be determined successively by substituting the expression for y_{n-1} obtained in the previous step into the right side of equation (10–61). Thus, substituting equation (10–60) into equation (10–61) with $n = 1$ gives

---

[105] Which is frequently referred to as the *boundary layer* region.

$$y_1(x) = -ae^{-\Omega(x)} \int_1^x \frac{1}{p(x)} \left[r^2(x) - r'(x)\right] dx \qquad (10\text{–}62)$$

where we have put

$$r(x) \equiv \frac{q(x)}{p(x)} \qquad (10\text{–}63)$$

and we shall suppose that the integral

$$A \equiv \int_1^0 \frac{1}{p(x)} \left[r^2(x) - r'(x)\right] dx \qquad (10\text{–}64)$$

exists.

Unlike the situation which occurred at the transition point, the zeroth-order solution remains bounded at $x=0$. In fact, its limiting value is $ae^{-\Omega(0)}$. But in order that the boundary condition (10–47) be satisfied, it is necessary that the asymptotic solution change from the value 1 to the value $ae^{-\Omega(0)}$ across the boundary layer region. Now we anticipate that the expansion (10–51) will hold over most of the region $0 \leq x \leq 1$ but that it will break down in a narrow region near $x=0$ whose thickness approaches zero as $\epsilon \rightarrow 0$. Since the asymptotic solution y must change from $ae^{-\Omega(0)}$ to 1 across this very narrow region, the derivatives y' and y'' must become very steep (large) in this region. However, in assuming that the solution had an asymptotic expansion of the form (10–51), we were essentially treating the terms y'', py', and qy in equation (10–46) as if they were of order 1. In order to overcome this difficulty, we proceed just as in the case of the turning point and rescale equation (10–46) in a manner which is appropriate to this boundary region by introducing a new "stretched" independent variable $\bar{x}$ by

$$\bar{x} = \frac{x}{\phi(\epsilon)} \qquad (10\text{–}65)$$

where $\lim_{\epsilon \to 0} \phi(\epsilon) = 0$ and the function ϕ is to be chosen so that it is of the scale of the boundary layer region. In this case, however, it is necessary to determine the scaling function ϕ in a different manner than in the case of the turning point. Thus, upon substituting equation (10–65) into equations (10–46) and (10–49), we obtain

$$\frac{\epsilon}{\phi^2(\epsilon)} \frac{d^2\bar{y}}{d\bar{x}^2} + \frac{p(\phi(\epsilon)\bar{x})}{\phi(\epsilon)} \frac{d\bar{y}}{d\bar{x}} + q(\phi(\epsilon)\bar{x})\bar{y} = 0 \qquad (10\text{--}66)$$

where we have put $\bar{y}(\bar{x}; \epsilon) = y$ and, in view of equations (10–49) and (10–50),

$$\left.\begin{aligned} p(\phi(\epsilon)\bar{x}) &= p_0 + p_1\phi(\epsilon)\bar{x} + p_2\phi^2(\epsilon)\bar{x}^2 + \cdots \\ q(\phi(\epsilon)\bar{x}) &= q_0 + q_1\phi(\epsilon)\bar{x} + q_2\phi^2(\epsilon)\bar{x}^2 + \cdots \end{aligned}\right\} \quad \text{for } p_0 \neq 0$$

$$(10\text{--}67)$$

Now in order that the inner solution satisfy the boundary condition at $x = 0$ and still contain another constant which can be adjusted to match the outer solution, we must require that the equation which is satisfied by the lowest order term in the expansion of the solution in the inner region (the inner expansion) be of second order. This equation is obtained by holding $\bar{x}$ and $\bar{y}$ fixed and taking the limit[106] $\epsilon \to 0$ in equation (10–66). However, the limiting form of this equation depends on the choice of the function $\phi(\epsilon)$. And in order that the highest order derivative be retained in this equation we must require that

$$\phi = O(\epsilon) \qquad (10\text{--}68)$$

If the condition $\phi = o(\epsilon)$ also held, the second and third terms of the limiting equation (10–66) would drop out and we would be left with the equation

$$\frac{\epsilon}{\phi^2(\epsilon)} \frac{d^2\bar{y}}{d\bar{x}^2} = 0 \qquad (10\text{--}69)$$

However, if we require that

$$\phi \neq o(\epsilon) \qquad (10\text{--}70)$$

then both the first and second terms will be retained in the limiting form of equation (10–66). In addition, the size of the inner region (which is determined by ϕ) will be larger if condition (10–70) holds than if it did not hold. Thus, the

---

[106] This limiting process is called the inner limit. The limit taken while holding x and y fixed and letting $\epsilon \to 0$ is called the outer limit.

solution which would be obtained by assuming that condition (10–70) did not hold is a limiting case (for small $\bar{x}$) of the inner solution which is obtained when equation (10–70) does hold. We shall, therefore, assume that conditions (10–68) and (10–70) hold. And since only the order of magnitude of ϕ is important, we will lose no generality by putting $\phi(\epsilon) = \epsilon$. Hence, equation (10–66) becomes, upon inserting equation (10–67),

$$\frac{d^2\bar{y}}{d\bar{x}^2} + (p_0 + \epsilon p_1 \bar{x} + \epsilon^2 p_2 \bar{x}^2 + \ldots) \frac{d\bar{y}}{d\bar{x}} + (\epsilon q_0 + \epsilon^2 q_1 \bar{x} + \ldots) \bar{y} = 0 \quad (10\text{–}71)$$

We now suppose that the solution to equation (10–71) has an asymptotic expansion of the form [107]

$$\bar{y}(\bar{x}; \epsilon) \sim \sum_{n=0}^{\infty} \bar{y}_n(\bar{x}) \epsilon^n \quad (10\text{–}72)$$

Then upon substituting this into equation (10–71) and equating to zero the coefficients of like powers of ϵ, we obtain

$$\frac{d^2\bar{y}_0}{d\bar{x}^2} + p_0 \frac{d\bar{y}_0}{d\bar{x}} = 0 \quad (10\text{–}73)$$

$$\frac{d^2\bar{y}_n}{d\bar{x}^2} + p_0 \frac{d\bar{y}_n}{d\bar{x}} = \bar{H}_n(\bar{x}) \qquad \text{for } n = 1, 2, \ldots \quad (10\text{–}74)$$

where we have put

$$\bar{H}_n(\bar{x}) = -\sum_{k=0}^{n-1} \left(\frac{d\bar{y}_k}{d\bar{x}} \bar{x} p_{n-k} + \bar{y}_k q_{n-k-1} \right) \bar{x}^{n-k-1} \qquad \text{for } n = 1, 2, \ldots$$

$$(10\text{–}75)$$

And upon substituting equation (10–72) into the boundary condition (10–47), we find that

$$\bar{y}_0(0) = 1 \quad (10\text{–}76)$$

---

[107] It should not be concluded from the results obtained so far that the asymptotic expansions of the solutions to equations containing a small parameter will always be power series in the parameter. For example, logarithmic terms could occur.

$$\bar{y}_n(0) = 0 \qquad \text{for } n = 1, 2, \ldots \qquad (10\text{--}77)$$

Since equations (10–73) and (10–74) do not contain the dependent variable explicitly, they can easily be integrated by the methods of chapter 4 to obtain

$$\bar{y}_0 = \bar{c}_0 + \bar{b}_0 e^{-p_0 \bar{x}}$$

$$\bar{y}_n = \bar{c}_n + \bar{b}_n e^{-p_0 \bar{x}} + \frac{1}{p_0} \int_0^{\bar{x}} [1 - e^{p_0(\bar{y} - \bar{x})}] \bar{H}_n(\bar{y}) \, d\bar{y} \qquad \text{for } n = 1, 2, \ldots$$

which become, after applying the boundary conditions (10–76) and (10–77),

$$\bar{y}_0 = \bar{c}_0 + (1 - \bar{c}_0) e^{-p_0 \bar{x}} \qquad (10\text{--}78)$$

$$\bar{y}_n = \bar{c}_n (1 - e^{-p_0 \bar{x}}) + \frac{1}{p_0} \int_0^{\bar{x}} [1 - e^{p_0(\bar{y} - \bar{x})}] \bar{H}_n(\bar{y}) \, d\bar{y} \qquad \text{for } n = 1, 2, \ldots$$
$$(10\text{--}79)$$

We can again determine the functions $\bar{y}_n$ for $n = 1, 2, \ldots$ successively by substituting the expressions for $\bar{y}_{n-1}$ obtained in the previous step into the right side of equation (10–79). Thus, upon substituting equation (10–78) into the right side of equation (10–79) with $n = 1$ and carrying out the integration, we obtain

$$\bar{y}_1 = \tilde{c}_1 (1 - e^{-p_0 \bar{x}}) + \frac{(1 - \bar{c}_0)}{p_0} \bar{x} e^{-p_0 \bar{x}} \left(q_0 - p_1 - \frac{p_1 p_0}{2} \bar{x} \right) - \bar{c}_0 \frac{q_0}{p_0} \bar{x}$$
$$(10\text{--}80)$$

where we have put

$$\tilde{c}_1 = \bar{c}_1 + \frac{1}{p_0^2} \left[(1 - \bar{c}_0)(p_1 - q_0) + \bar{c}_0 q_0 \right]$$

DIFFERENTIAL EQUATIONS

As in the case of the transition point, we now suppose that there is an overlap domain in which the outer expansion (10–51) and the inner expansion (10–72) both apply and that within this overlap domain, these two expansions agree to within an error which is of smaller order than the last term retained. In order to perform this matching, we introduce an intermediate variable

$$x^+ = \frac{x}{\nu(\epsilon)} = \frac{\epsilon \bar{x}}{\nu(\epsilon)} \tag{10–81}$$

into both the inner and outer expansions. The scale factor $\nu(\epsilon)$ is determined so that the variable x^+ is of order 1 in the overlap domain. Since the size of the overlap domain must approach zero as $\epsilon \to 0$ and since it must be farther away from $x = 0$ than the inner regions whose size is $O(\epsilon)$, we must require that

$$\epsilon = o(\nu) \tag{10–82}$$

and

$$\nu = o(1) \tag{10–83}$$

To proceed with the matching, we then reexpand the first m terms for $m = 1, 2,$ . . . of both the inner and outer expansions while holding x^+ fixed and then require that the difference between these two expansions be of $o(\epsilon^{m-1})$. That is, the inner and outer expansions are asymptotically equal in the overlap domain to the appropriate order. Hence, we require that

$$y_0[\nu(\epsilon)x^+] + \epsilon y_1[\nu(\epsilon)x^+] + \ldots + \epsilon^n y_n[\nu(\epsilon)x^+] - \bar{y}_0\left[\frac{\nu(\epsilon)}{\epsilon}x^+\right]$$

$$- \epsilon \bar{y}_1\left[\frac{\nu(\epsilon)}{\epsilon}x^+\right] - \ldots - \epsilon^n \bar{y}_n\left[\frac{\nu(\epsilon)}{\epsilon}x^+\right] = o(\epsilon^n) \qquad \text{as } \epsilon \to 0,$$

$$\text{for } n = 0, 1, 2, \ldots \text{ with } x^+ \text{ fixed} \tag{10–84}$$

Thus, for example, when $n = 0$, this becomes

$$y_0[\nu(\epsilon)x^+] - \bar{y}_0\left[\frac{\nu(\epsilon)}{\epsilon}x^+\right] = o(1) \qquad \text{as } \epsilon \to 0 \text{ with } x^+ \text{ fixed} \tag{10–85}$$

And when $n = 1$, it becomes

$$y_0 \left[\nu(\epsilon) x^+ \right] + \epsilon y_1 \left[\nu(\epsilon) x^+ \right] - \bar{y}_0 \left[\frac{\nu(\epsilon)}{\epsilon} x^+ \right] - \epsilon \bar{y}_1 \left[\frac{\nu(\epsilon)}{\epsilon} x^+ \right] = o(\epsilon)$$

$$\text{as } \epsilon \to 0 \text{ with } x^+ \text{ fixed} \qquad (10\text{--}86)$$

Now since

$$\Omega[\nu(\epsilon) x^+] = \int_1^{\nu x^+} \frac{q(\nu x^+)}{p(\nu x^+)} \, \nu \, dx^+ = \int_1^{\nu x^+} \frac{q(\xi)}{p(\xi)} \, d\xi = \int_1^0 \frac{q(\xi)}{p(\xi)} \, d\xi$$

$$+ \int_0^{\nu x^+} \frac{q(\xi)}{p(\xi)} \, d\xi = \Omega(0) + \nu(\epsilon) x^+ \frac{q_0}{p_0} + O(\nu^2)$$

it follows, upon expanding the exponential, that

$$e^{-\Omega[\nu(\epsilon)x^+]} = e^{-\Omega(0)} e^{-\nu x^+(q_0/p_0) +} O(\nu^2) = e^{-\Omega(0)} \left[1 - \nu(\epsilon) \frac{q_0}{p_0} x^+ + O(\nu^2) \right]$$

And, the solutions (10–60) and (10–62) therefore become, respectively,

$$y_0[\nu(\epsilon)x^+] = a e^{-\Omega(0)} \left[1 - \nu(\epsilon) \frac{q_0}{p_0} x^+ + O(\nu^2) \right] \qquad (10\text{--}87)$$

$$y_1 [\nu(\epsilon) x^+] = a e^{-\Omega(0)} [A + O(\nu)] \qquad (10\text{--}88)$$

where definition (10–64) has been used. On the other hand, the solution (10–78) becomes

$$\bar{y}_0 \left[\frac{\nu(\epsilon)}{\epsilon} x^+ \right] = \bar{c}_0 + (1 - \bar{c}_0) e^{-p_0[\nu(\epsilon)/\epsilon] x^+}$$

But, since equation (10–82) shows that

303

$$\lim_{\epsilon \to 0} \frac{\nu(\epsilon)}{\epsilon} = \infty$$

this becomes

$$\bar{y}_0 \left[\frac{\nu(\epsilon)}{\epsilon} x^+ \right] = \bar{c}_0 + \text{exponentially small terms} \qquad (10\text{--}89)$$

Similarly, the solution (10–80) becomes

$$\bar{y}_1 \left[\frac{\nu(\epsilon)}{\epsilon} x^+ \right] = \tilde{c}_1 - \bar{c}_0 \frac{q_0}{p_0} \frac{\nu(\epsilon)}{\epsilon} x^+ + \text{exponentially small terms}$$

$$(10\text{--}90)$$

Equations (10–87) and (10–89) now show that

$$y_0[\nu(\epsilon)x^+] - \bar{y}_0 \left[\frac{\nu(\epsilon)}{\epsilon} x^+ \right] = ae^{-\Omega(0)} - \bar{c}_0 + O(\nu)$$

since the exponentially small terms are of lower order than any power of ν. But since $\lim_{\epsilon \to 0} \nu(\epsilon) = 0$, this shows that the zeroth-order matching condition (10–85) will be satisfied provided that

$$\bar{c}_0 = ae^{-\Omega(0)} \qquad (10\text{--}91)$$

Similarly, equations (10–86) to (10–90) show upon substituting in equation (10–91) that

$$y_0[\nu(\epsilon)x^+] + \epsilon y_1[\nu(\epsilon)x^+] - \bar{y}_0 \left[\frac{\nu(\epsilon)}{\epsilon} x^+ \right] - \epsilon \bar{y}_1 \left[\frac{\nu(\epsilon)}{\epsilon} x^+ \right]'$$

$$= ae^{-\Omega(0)} \left(1 - \nu \frac{q_0}{q_1} x^+ + \epsilon A \right) - ae^{-\Omega(0)} \left(1 - \nu \frac{q_0}{q_1} x^+ \right) - \epsilon \tilde{c}_1 + O(\nu^2)$$

$$= \epsilon [e^{-\Omega(0)} aA - \tilde{c}_1] + O(\nu^2)$$

Hence, the first-order matching condition (10–86) will now be satisfied if we put $\tilde{c}_1 = ae^{-\Omega(0)}A$ and choose the scaling function ν so that

$$\nu^2 = o(\epsilon) \qquad (10\text{–}92)$$

Notice that ν must simultaneously satisfy conditions (10–82), (10–83), and (10–92). There are many choices of ν which will accomplish this. For example, we can take $\nu(\epsilon) = \epsilon^{2/3}$.

We have now obtained the following results: In the outer region, the expansion to two terms is

$$y(x; \epsilon) \sim ae^{-\Omega(x)}\left\{1 + \epsilon\int_1^x \frac{1}{p(x)}\left[r^2(x) - r'(x)\right]dx + \ldots\right\} \qquad (10\text{–}93)$$

In the inner region, the expansion to two terms is

$$\bar{y}(\bar{x}; \epsilon) \sim e^{-p_0\bar{x}}\left\{\left[1 - ae^{-\Omega(0)}\right]\left[1 + \frac{\epsilon\bar{x}}{p_0}\left(q_0 - p_1 - \frac{p_1 p_0 \bar{x}}{2}\right)\right] - ae^{-\Omega(0)}\epsilon A\right\}$$

$$+ ae^{-\Omega(0)}\left(1 + \epsilon A - \frac{q_0}{p_0}\epsilon\bar{x}\right) + \ldots \qquad (10\text{–}94)$$

And in the overlap domain, both of these expansions take on the common value y_t given by

$$y_t \sim ae^{-\Omega(0)}\left(1 + \epsilon A - \frac{q_0}{p_0}\epsilon\bar{x}\right) \qquad (10\text{–}95)$$

As in the case of the transition point, we can obtain an expansion which is valid everywhere in the interval $0 \leqslant x \leqslant 1$ (that is, a uniformly valid expansion) by adding the inner expansion to the outer expansion and, so that it will not be counted twice, subtracting their common value in the overlap domain. Thus,

$$y_{\text{uniformly valid}} \sim ae^{-\Omega(x)}\left\{1 + \epsilon\int_1^x \frac{1}{p(x)}\left[r^2(x) - r'(x)\right]dx\right\}$$

$$+ e^{-p_0\bar{x}}\left\{\left[1 - ae^{-\Omega(0)}\right]\left[1 + \frac{\epsilon\bar{x}}{p_0}\left(q_0 - p_1 - \frac{p_1 p_0}{2}\bar{x}\right)\right] - ae^{-\Omega(0)}\epsilon A\right\} + \ldots \qquad (10\text{–}96)$$

Of course, we have not actually proved that equation (10−96) is indeed an asymptotic expansion of the solution to equation (10−46) subject to the boundary conditions (10−47) and (10−48). However, it is common practice in applied mathematics to accept an expansion obtained by a formal procedure (such as those given in this section) as being a true asymptotic expansion of the solution to the problem. Indeed, most problems which are treated in practice are too complicated for rigorous proofs to be carried through.

We emphasize again that this method applies to nonlinear equations just as well as to linear equations.

10.4 METHOD OF STRAINED COORDINATES

We shall now consider some additional techniques which can be used to handle the nonuniformities that arise when asymptotic solutions are sought to certain types of differential equations. Thus, consider the differential equation

$$\frac{d^2y}{dt^2} + \omega^2 y = \epsilon y^3 \tag{10−97}$$

which describes the oscillation of a mass on a spring with a weakly nonlinear restoring force. And for definiteness, let us impose the initial conditions

$$y(0) = 1 \qquad \frac{dy}{dt}(0) = 0 \tag{10−98}$$

If we attempt to find an asymptotic solution in the form of a straight-forward power series in ϵ

$$y(t; \epsilon) \sim y_0(t) + \epsilon y_1(t) + \ldots$$

then we get, upon substituting this into equations (10−97) and (10−98) and equating to zero the coefficients of like powers of ϵ,

$$\frac{d^2y_0}{dt^2} + \omega^2 y_0 = 0 \tag{10−99}$$

$$\frac{d^2 y_1}{dt^2} + \omega^2 y_1 = y_0^3 \tag{10-100}$$

$$y_0(0) = 1 \qquad \frac{dy_0}{dt}(0) = 0 \tag{10-101}$$

$$y_1(0) = \frac{dy_1}{dt}(0) = 0 \tag{10-102}$$

However, the solution to equation (10–99) subject to the initial conditions (10–101) is $y_0 = \cos \omega t$. And after substituting this into equation (10–100), we obtain

$$\frac{d^2 y_1}{dt^2} + \omega^2 y_1 = \cos^3 \omega t = \frac{3}{4} \cos \omega t + \frac{1}{4} \cos 3\omega t$$

But since the general solution to this equation is

$$y_1 = -\frac{1}{32\omega^2} \cos 3\omega t + \frac{3}{8} \frac{t}{\omega} \sin \omega t + c_1 \cos \omega t + c_2 \sin \omega t$$

the asymptotic solution to equation (10–97) is

$$y \sim \cos \omega t + \epsilon \left[\frac{3}{8} \frac{t}{\omega} \sin \omega t - \frac{1}{32\omega^2} \cos 3\omega t + c_1 \cos \omega t + c_2 \sin \omega t \right] + \dots$$

However, the first-order term in this expansion will not be small compared with the zeroth-order term when $t \approx 1/\epsilon$ due to the presence of the term $t \sin \omega t$ in the solution. Such terms are called *secular terms*. The expansion will, therefore, not be valid for large times even though it may be suitable for calculating the solution at small times. The solution of equation (10–97) is periodic. However, due to the appearance of secular terms, the solution cannot be carried to sufficiently long times to calculate the distortion of the period due to the nonlinear restoring force.

A method (based on the work of Poincaré[108]) for alleviating this difficulty

---

[108] The ideas were developed in the course of his work on periodic orbits of the planets.

was developed by Whitham and by Lighthill (ref. 37). The method depends upon introducing a new variable τ and then considering both the dependent variable y and the independent variable t to be functions of τ. Thus,

$$t = t(\tau;\ \epsilon)$$

$$y = y(\tau;\ \epsilon)$$

And we suppose that these functions can be expanded in powers of ϵ to obtain

$$t = \tau + \epsilon t_1(\tau) + \ldots \tag{10-103}$$

$$y = y_0(\tau) + \epsilon y_1(\tau) + \ldots \tag{10-104}$$

The arbitrariness introduced into the solution by the functions t_1, etc., is used to adjust the nonhomogeneous terms in the equations for $y_1, y_2, \ldots$ so that the secular terms will not occur in the expansion. The resulting distortion of the time scale will cause the frequency of the motion to depend on ϵ. And this will allow us to calculate the variation of the period.

Now it follows from equation (10–103) that

$$\frac{d}{dt} = \frac{d\tau}{dt}\frac{d}{d\tau} = \frac{1}{(1+\epsilon t_1' + \ldots)}\frac{d}{d\tau} = (1 - \epsilon t_1' + \ldots)\frac{d}{d\tau} \tag{10-105}$$

where the prime denotes differentiation with respect to τ. Hence,

$$\frac{d^2 y}{d\tau^2} = (1 - \epsilon t_1' - \ldots)\frac{d}{d\tau}\left[(1 - \epsilon t_1' - \ldots)\left(\frac{dy_0}{d\tau} + \epsilon\frac{dy_1}{d\tau} + \ldots\right)\right]$$

$$= (1 - 2\epsilon t_1' + \ldots)\frac{d^2 y_0}{d\tau^2} + \epsilon\frac{d^2 y_1}{d\tau^2} - \epsilon t_1''\frac{dy_0}{d\tau} + \ldots$$

After substituting this into equation (10–97) together with the expansion (10–104) and equating to zero the coefficients of like powers of ϵ, we obtain

$$\frac{d^2 y_0}{d\tau^2} + \omega^2 y_0 = 0 \tag{10-106}$$

$$\frac{d^2y_1}{d\tau^2} + \omega^2 y_1 = y_0^3 + 2t_1' \frac{d^2y_0}{d\tau^2} + t_1'' \frac{dy_0}{d\tau} \qquad (10\text{--}107)$$

.

.

.

In order to find the proper boundary conditions, we need expressions for the values of y and dy/dt at $t = 0$ in terms of their values at $\tau = 0$. Thus, we use a Taylor series expansion of these quantities about $\tau = 0$ to obtain

$$y(t = 0) = y|_{\tau=0} + \left(\frac{dy}{d\tau}\right)_{\tau=0} \tau(0) + \ldots$$

$$\frac{dy}{dt}\bigg|_{t=0} = \left(\frac{dy}{dt}\right)_{\tau=0} + \left(\frac{d^2y}{d\tau dt}\right)_{\tau=0} \tau(0) + \ldots$$

where $\tau(0)$ denotes the value of the function $\tau(t)$ (obtained by solving equation (10–103) for τ as a function of t) at the point $t = 0$. Upon substituting in equations (10–103) to (10–105), we find

$$y(t = 0) = y_0(0) + \epsilon \left[y_1(0) + t_1(0) \left(\frac{dy_0}{d\tau}\right)_{\tau=0} \right] + \ldots$$

$$\frac{dy}{dt}\bigg|_{t=0} = \left(\frac{dy_0}{d\tau}\right)_{\tau=0} + \epsilon \left[\left(\frac{dy_1}{d\tau}\right)_{\tau=0} - t_1'(0) \left(\frac{dy_0}{d\tau}\right)_{\tau=0} + t_1(0) \left(\frac{d^2y_0}{d\tau^2}\right)_{\tau=0} \right] + \ldots$$

Then substituting these expressions into the boundary conditions (10–98) and equating to zero the coefficients of like powers of ϵ shows that

$$y_0(0) = 1 \qquad \frac{dy_0}{d\tau}(0) = 0 \qquad (10\text{--}108)$$

$$y_1(0) + t_1(0) \frac{dy_0}{d\tau}(0) = 0$$

$$\frac{dy_1}{d\tau}(0) - t_1'(0)\frac{dy_0}{d\tau}(0) + t_1(0)\frac{d^2y_0}{d\tau^2}(0) = 0$$

or, using the boundary condition (10–108) to simplify the second group of boundary conditions,

$$\left.\begin{array}{c} y_1(0) = 0 \\[2mm] \dfrac{dy_1}{d\tau}(0) + t_1(0)\dfrac{d^2y_0}{d\tau^2}(0) = 0 \end{array}\right\} \qquad (10\text{–}109)$$

The solution to equation (10–106) subject to the boundary conditions (10–108) is

$$y_0 = \cos \omega\tau \qquad (10\text{–}110)$$

This has the same form as the zeroth-order solution obtained by the regular perturbation procedure. However, upon substituting equation (10–110) into equation (10–107), the latter equation becomes

$$\frac{d^2y_1}{d\tau^2} + \omega^2 y_1 = \cos^3 \omega\tau - 2\omega^2 t_1' \cos \omega\tau - \omega t_1'' \sin \omega\tau$$

$$= \frac{1}{4}\cos 3\omega\tau + \left(\frac{3}{4} - 2\omega^2 t_1'\right)\cos \omega\tau - \omega t_1'' \sin \omega\tau$$

We now determine the function $t_1(\tau)$ so that secular terms will not appear in the first-order solution y_1. These terms arise because the solutions $\cos \omega\tau$ and $\sin \omega\tau$ of the homogeneous equation appear in the nonhomogeneous terms. Thus, the secular terms will not occur if we choose t_1 so that the coefficients of these two terms vanish. Hence, we take $t_1'' = 0$, $t_1' = 3/8\omega^2$, or

$$t_1 = \frac{3}{8\omega^2}\,\tau \qquad (10\text{–}111)$$

where we have set the constant of integration equal to zero in order to simplify the second boundary condition (10–109). And the differential equation for y_1 becomes

$$\frac{d^2 y_1}{d\tau^2} + \omega^2 y_1 = \frac{1}{4} \cos 3\omega\tau$$

Now since substituting equation (10–111) into the expansion (10–103) shows that

$$t = \tau + \frac{3}{8\omega^2} \epsilon\tau + \ldots$$

we find upon neglecting higher order terms in ϵ that

$$\tau = \frac{1}{\left(1 + \dfrac{3\epsilon}{8\omega^2}\right)} t + \ldots = \left(1 - \frac{3\epsilon}{8\omega^2}\right) t + \ldots$$

And when this is substituted into the zeroth-order solution (10–110), we obtain

$$y_0 = \cos\left(\omega - \frac{3\epsilon}{8\omega}\right) t$$

Thus, we find that the frequency is modified by the nonlinear restoring force even in the lowest order term of this expansion. In contrast to this, the direct expansion leads to no information about the frequency change. The expansion can be carried to higher orders by continuing to use the functions t_n for $n = 2, 3, \ldots$ to eliminate the secular terms. And this expansion will be uniformly valid for all times since no secular terms will appear.

10.5 MULTIPLE-TIME METHODS

Equations which contain two disparate time or length scales are frequently encountered in practice. This happens, for example, in problems which involve a small force acting over a long period of time. An example of this is a spring-mass system subjected to a weak viscous damping. The two times which occur in this problem are the basic period of oscillation and the damping time which occurs over many periods. The basic idea of the method is due to Kuzmark.

And a detailed discussion of this method is given by Kevorkian in reference 38. The presentation and examples used herein are taken from this report. The method takes into account the long-time effects in such a way as to render the expansion uniformly valid. As in the preceding example, it involves reasoning about the first-order terms to determine the zeroth-order solution.

We set up the expansion in such a way that a fast time variable t and a slow time variable $\tilde{t}$ are explicitly exhibited. The slow variable is assumed to be related to the fast variable by

$$\tilde{t} = \phi(\epsilon)t \tag{10-112}$$

where $\lim_{\epsilon \to 0} \phi(\epsilon) = 0$. And we suppose that the solution to the differential equation has an expansion of the form

$$y(t; \epsilon) \sim \nu_0(\epsilon)f_0(t^*, \tilde{t}) + \nu_1(\epsilon)f_1(t^*, \tilde{t}) + \nu_2(\epsilon)f_2(t^*, \tilde{t}) + \ldots \tag{10-113}$$

where $\{\nu_j(\epsilon)\}$ is an asymptotic sequence as $\epsilon \to 0$ and t^* is another fast time variable related to the fast variable t in such a way that it accounts for the change in frequency that occurs in the problem. This relation is taken in the form

$$t^* = t[1 + \mu_1(\epsilon)\omega_1 + \mu_2(\epsilon)\omega_2 + \ldots] \tag{10-114}$$

where $\{\mu_j(\epsilon)\}$ is another asymptotic sequence and $\{\omega_j\}$ is a sequence of constants which are determined by the problem in such a way as to render the expansion (10–113) uniformly valid.

The variables t^* and $\tilde{t}$ in equation (10–113) will be treated as independent. Hence, the ordinary differential equation satisfied by y will be transformed into a sequence of partial differential equations. However, it turns out that they can still be treated as ordinary differential equations.

Upon differentiating equation (10–113) with respect to t, we find that

$$\frac{dy}{dt} = \left(\frac{\partial y}{\partial t^*}\right)_{\tilde{t}} \frac{dt^*}{dt} + \left(\frac{\partial y}{\partial \tilde{t}}\right)_{t^*} \frac{d\tilde{t}}{dt}$$

But since equations (10–112) and (10–114) show that

$$\frac{dt^*}{dt} = 1 + \mu_1(\epsilon)\omega_1 + \mu_2(\epsilon)\omega_2 + \ . \ . \ .$$

$$\frac{d\tilde{t}}{dt} = \phi(\epsilon)$$

this becomes

$$\frac{dy}{dt} = [1 + \mu_1(\epsilon)\omega_1 + \ . \ . \ .] \left[\nu_0(\epsilon)\frac{\partial f_0}{\partial t^*} + \nu_1(\epsilon)\frac{\partial f_1}{\partial t^*} + \ . \ . \ . \right]$$

$$+ \phi(\epsilon) \left[\nu_0(\epsilon)\frac{\partial f_0}{\partial \tilde{t}} + \nu_1(\epsilon)\frac{\partial f_1}{\partial \tilde{t}} + \ . \ . \ . \right]$$

or

$$\frac{dy}{dt} = \nu_0(\epsilon) \left\{ \frac{\partial f_0}{\partial t^*} [1 + \mu_1(\epsilon)\omega_1] + \phi(\epsilon)\frac{\partial f_0}{\partial \tilde{t}} \right\} + \nu_1(\epsilon)\frac{\partial f_1}{\partial t^*} + \ . \ . \ .$$

This equation shows that changes which occur on the slow time scale $\tilde{t}$ are small compared with the changes which occur on the fast time scale.

The method is best illustrated by considering a particular example. Thus, consider the equation

$$\frac{d^2y}{dt^2} + y + \epsilon \left(\frac{dy}{dt} \right)^3 = 0 \tag{10--115}$$

which governs the behavior of an oscillator with weak cubic damping. We shall seek an asymptotic solution which is uniformly valid for $0 \le t \le \infty$, subject to the boundary conditions

$$y(0) = 0 \tag{10--116}$$

$$\frac{dy}{dt}(0) = 1 \tag{10--117}$$

In this case, it is reasonable to begin by choosing the functions ϕ and the two asymptotic sequences $\{\nu_j\}$ and $\{\mu_j\}$ to be $\phi(\epsilon) = \epsilon$ and $\nu_j = \mu_j = \epsilon^j$. Then

$\tilde{t} = \epsilon t$ and the expansion (10–113) becomes

$$y\,(t;\,\epsilon) = f_0\,(t^*, \tilde{t}\,) + \epsilon f_1\,(t^*, \tilde{t}\,) + \ldots$$

with

$$t^* = t\,(1 + \omega_2 \epsilon^2 + \ldots)$$

where we have omitted the term $\epsilon \omega_1 t$ since we wish to ensure that ϵt only occurs in the solution [109] as $\tilde{t}$. Hence,

$$\frac{dy}{dt} = \frac{\partial f_0}{\partial t^*} + \epsilon \left(\frac{\partial f_0}{\partial \tilde{t}} + \frac{\partial f_1}{\partial t^*} \right) + O(\epsilon^2)$$

$$\frac{d^2 y}{dt^2} = \frac{\partial^2 f_0}{\partial t^{*2}} + 2\epsilon \frac{\partial^2 f_0}{\partial t^* \partial \tilde{t}} + \epsilon \frac{\partial^2 f_1}{\partial t^{*2}} + O(\epsilon^2)$$

Upon substituting these results into equation (10–115) and equating to zero the coefficients of like powers of ϵ, we find that

$$\frac{\partial^2 f_0}{\partial t^{*2}} + f_0 = 0 \tag{10–118}$$

$$\frac{\partial^2 f_1}{\partial t^{*2}} + 2 \frac{\partial^2 f_0}{\partial t^* \partial t} + \left(\frac{\partial f_0}{\partial t^*} \right)^3 + f_1 = 0 \tag{10–119}$$

.

.

.

And since $y(0) = f_0(0, 0) + \epsilon f_1(0, 0) + \ldots$ and

---

[109] There is a certain amount of arbitrariness in the choice of the variables t^* and $\tilde{t}$.

$$\frac{dy}{dt}(0) = \frac{\partial f_0}{\partial t^*}\bigg|_{0,0} + \epsilon\left(\frac{\partial f_0}{\partial \tilde{t}}\bigg|_{0,0} + \frac{\partial f_1}{\partial t^*}\bigg|_{0,0}\right) + \cdots$$

the boundary conditions (10–116) and (10–117) show that

$$f_0(0,0) = 0 \qquad \frac{\partial f_0}{\partial t^*}\bigg|_{0,0} = 1 \qquad\qquad (10\text{–}120)$$

$$f_1(0,0) = 0 \qquad \frac{\partial f_1}{\partial t^*}\bigg|_{0,0} + \frac{\partial f_0}{\partial \tilde{t}}\bigg|_{0,0} = 0 \qquad\qquad (10\text{–}121)$$

.

.

.

Now the solution to equation (10–118) is

$$f_0(t^*, \tilde{t}) = C_0(\tilde{t})\sin t^* + D_0(\tilde{t})\cos t^* \qquad\qquad (10\text{–}122)$$

where the functions $C_0(\tilde{t})$ and $D_0(\tilde{t})$ are arbitrary functions arising from the integration. But substituting equation (10–122) into the boundary conditions (10–120) shows that

$$D_0(0) = 0 \qquad C_0(0) = 1 \qquad\qquad (10\text{–}123)$$

As in the preceding method, we now determine the functions C_0 and D_0 so that the first-order solution f_1 will not contain any secular terms. To this end, notice that equation (10–122) implies that

$$\left(\frac{\partial f_0}{\partial t^*}\right)^3 = (C_0\cos t^* - D_0\sin t^*)^3 = C_0^3\cos^3 t^* - 3D_0 C_0^2\sin t^*\cos^2 t^*$$

$$+ 3D_0^2 C_0\sin^2 t^*\cos t^* - D_0^3\sin^3 t^*$$

Hence, we find, upon using the identities

315

$$\cos^3 t^* = \frac{1}{4} \cos 3t^* + \frac{3}{4} \cos t^*$$

$$\sin^3 t^* = -\frac{1}{4} \sin 3t^* + \frac{3}{4} \sin t^*$$

$$\sin t^* \cos^2 t^* = \sin t^* - \sin^3 t^*$$

$$\sin^2 t^* \cos t^* = \cos t^* - \cos^3 t^*$$

that

$$\left(\frac{\partial f_0}{\partial t^*}\right)^3 = \frac{3}{4} C_0(C_0^2 + D_0^2) \cos t^* - \frac{3}{4} D_0(C_0^2 + D_0^2) \sin t^*$$

$$+ \frac{1}{4} D_0(D_0^2 - 3C_0^2) \sin 3t^* + \frac{1}{4} C_0(C_0^2 - 3D_0^2) \cos 3t^* \qquad (10\text{--}124)$$

We therefore find, after substituting equations (10–122) and (10–124) into equation (10–119), that

$$\frac{\partial^2 f_1}{\partial t^{*2}} + f_1 = \left[\frac{3}{4} D_0(D_0^2 + C_0^2) + 2\frac{dD_0}{d\tilde{t}}\right] \sin t^* - \left[\frac{3}{4} C_0(D_0^2 + C_0^2) + 2\frac{dC_0}{d\tilde{t}}\right] \cos t^*$$

$$- \frac{1}{4} D_0(D_0^2 - 3C_0^2) \sin 3t^* - \frac{1}{4} C_0(C_0^2 - 3D_0^2) \cos 3t^* \qquad (10\text{--}125)$$

In order to ensure that secular terms do not occur in f_1, we must eliminate $\sin t^*$ and $\cos t^*$ from the nonhomogeneous term of this equation. This can be accomplished by putting

$$\left.\begin{aligned} 2\frac{dD_0}{d\tilde{t}} + \frac{3}{4} D_0(D_0^2 + C_0^2) &= 0 \\[2ex] 2\frac{dC_0}{d\tilde{t}} + \frac{3}{4} C_0(D_0^2 + C_0^2) &= 0 \end{aligned}\right\} \qquad (10\text{--}126)$$

Upon multiplying the first of these by D_0 and the second by C_0 and adding the results, we get

$$\frac{d}{d\tilde{t}}(D_0^2 + C_0^2) + \frac{3}{4}(D_0^2 + C_0^2)^2 = 0$$

which has the solution

$$D_0^2 + C_0^2 = \frac{1}{\frac{3}{4}\tilde{t} + K_1}$$

where K_1 is a constant. And substituting this into the boundary condition (10–123) shows that we must take $K_1 = 1$ to obtain

$$D_0^2 + C_0^2 = \frac{1}{\frac{3}{4}\tilde{t} + 1} \tag{10–127}$$

But substituting this into the first equation (10–126) shows that

$$2\frac{dD_0}{d\tilde{t}} + \frac{3}{4}\frac{D_0}{\frac{3}{4}\tilde{t} + 1} = 0$$

Since the solution to this equation subject to the first boundary condition (10–123) is $D_0(\tilde{t}) = 0$, equation (10–127) becomes

$$C_0 = \frac{1}{\sqrt{\frac{3}{4}\tilde{t} + 1}}$$

where the positive square root is taken to ensure that C_0 satisfy the second boundary condition (10–123). Substituting these results into the zeroth-order solution (10–122) now shows that

$$f_0(t^*, \tilde{t}) = \frac{\sin t^*}{\sqrt{\frac{3}{4}\tilde{t} + 1}}$$

Hence, we obtain the one-term uniformly valid asymptotic expansion of the solution to equation (10−115)

$$y \sim \frac{\sin t^*}{\sqrt{\frac{3}{4}\tilde{t} + 1}} = \frac{\sin t}{\sqrt{1 + \frac{3}{4}\epsilon t}}$$

The procedure can be continued to obtain higher order terms, the solution to any given order being determined by reasoning about the next higher order terms.

CHAPTER 11

Numerical Methods

It frequently happens that the differential equations encountered in practice cannot be solved by the exact and approximate methods discussed in the preceding chapters. In such cases it is often necessary to resort to numerical methods in conjunction with a digital computer. These methods usually involve replacing the differential equations by a number of algebraic equations, called *difference equations*, in such a way that the solution of the difference equations is in some sense close to that of the differential equation. There are a large number of numerical procedures available. The choice of method is influenced by the type of auxiliary conditions as well as by the form of the equation. Thus, when all the auxiliary conditions are imposed at a single point (initial conditions), the solution can be developed by means of "marching techniques," which solve the difference equations in succession. These marching solutions can be carried out either by using implicit methods such as Euler's method and the Runge-Kutta method or by using explicit methods such as the Adams method and the improved Euler method. Each of these methods has advantages and disadvantages which will be discussed subsequently. When auxiliary conditions are imposed at two points (boundary conditions), it is usually possible to solve linear equations directly by using either the superposition principle or finite difference schemes which involve matrix methods. However, when the equations are nonlinear, it is often necessary either to linearize the problem or to reduce it to a set of initial-value problems and then use iterative or matrix techniques to obtain the solution.

The subject of numerical solutions to differential equations is quite vast and we cannot hope to cover it completely in a single chapter. For more detailed information, the reader is referred to references 39 to 42.

11.1 APPROXIMATION BY DIFFERENCE EQUATIONS: ERRORS AND INSTABILITY

Consider the general system of nth-order differential equations

$$\mathbf{F}(\mathbf{y}^{(n)}, \mathbf{y}^{(n-1)}, \ldots, \mathbf{y}, x) = 0 \tag{11-1}$$

defined on the interval $a \leq x \leq b$ and subject to appropriate initial or boundary conditions.

In order to obtain a numerical solution to this system we first *partition* the interval $a \leq x \leq b$. A partition of the interval $a \leq x \leq b$ is defined to be any finite set of points $x_1, x_2, \ldots, x_{m+1}$ which has the property that $a = x_1 < x_2 < \ldots < x_{m+1} = b$. The length h_j of the jth subinterval,[110] $x_j \leq x \leq x_{j+1}$, is called the *step size*. Thus,

$$h_j = x_{j+1} - x_j \qquad \text{for } j = 1, 2, \ldots, m$$

The system of differential equations (11-1) is then "replaced" by a set of algebraic equations, say

$$\mathbf{G}_k(\mathbf{y}_1, \ldots, \mathbf{y}_{m+1}; x_1, \ldots, x_{m+1}) = 0 \qquad \text{for } k = 1, 2, \ldots, p \tag{11-2}$$

called *difference equations*. Their solution is the set of $m + 1$ vectors $\mathbf{y}_1, \ldots, \mathbf{y}_{m+1}$, which are approximately equal to the values taken on by the solution $\mathbf{y} = \mathbf{f}(x)$ of the system (11-1) at the $m + 1$ points $x_1, \ldots, x_{m+1}$. Before discussing the various methods whereby such difference equations can be constructed, we shall first consider certain types of errors which can occur when difference equations are used to obtain numerical solutions.

The *discretization* or *truncation* error E_i at the ith step is defined to be the magnitude of the difference between the solution to the differential equations at the point x_i and the solution $\mathbf{y}_i$ of the difference equations. Thus

$$E_i = |\mathbf{f}(x_i) - \mathbf{y}_i|$$

This error depends only on the type of difference equation used and is independent of the method by which it is solved.

---

[110] It is not necessary to have all h_j equal, but it is usually desirable.

However, there is also an error which is caused by the numerical procedure itself. Because a computer can accommodate only a limited number of significant figures, it cannot store accurately an irrational number or even a rational number requiring precision beyond the computer's capability. Therefore, the computers themselves introduce an error which results from the necessity of rounding off numbers: it is known as the *round-off error*. The round-off error at any step in the computations propagates to the next step and is combined there with the round-off error of that step. The generation of the round-off error at each step is extremely unpredictable. Precisely for this reason, analyses of round-off error often treat the error per step as a random variable (see ref. 43).

Closely related to the question of error is the question of *stability*, which must be considered in certain instances before a numerical solution can be obtained. Various types of instability can arise. If the instability is inherent in the differential equation itself, it is called an *inherent instability*. For example, consider the initial-value problem

$$\left.\begin{array}{l} \dfrac{dy_1}{dx} = y_2 \\[2em] \dfrac{dy_2}{dx} = 100y_1 \end{array}\right\} \tag{11-3}$$

subject to the initial conditions $y_1(0) = 1$ and $y_2(0) = -10$. The general solution for equations (11-3)

$$y_1(x) = C_1 e^{-10x} + C_2 e^{10x}$$

$$y_2(x) = -10C_1 e^{-10x} + 10C_2 e^{10x}$$

And the initial conditions are satisfied by taking $C_1 = 1$ and $C_2 = 0$. However, when a numerical procedure is used, it will usually be impossible to satisfy these initial conditions exactly. But a small error in determining the constant C_2 will then allow the second terms in the solutions to eventually grow so large that they will dominate the first terms, which correspond to the solution being sought. Situations of this type arise most frequently when the initial-value problem is being solved as part of an iteration procedure to solve a boundary-value problem.

Instabilities can also arise from the difference equations (even when the differential equations are stable). They are then called *induced instabilities*. These instabilities can result in spurious solutions to the difference equations which do not correspond to solutions of the differential equations. For more details the reader is referred to reference 43.

11.2 INITIAL-VALUE PROBLEMS

In this section we shall consider certain types of "difference schemes" which are suitable for obtaining numerical solutions to initial-value problems.

11.2.1 Explicit Methods

11.2.1.1 *One-step processes: Taylor series method.*—It has been indicated in chapter 3 that any normal system of ordinary differential equations can always be written as a first-order normal system, which in vector notation has the form

$$\frac{d\mathbf{y}}{dx} = \mathbf{G}(x, \mathbf{y}) \tag{11-4}$$

We shall now consider some methods, referred to as *one-step processes*, for obtaining numerical solutions to this equation on an interval $a \leqslant x \leqslant b$ subject to the initial conditions $\mathbf{y} = \mathbf{y}_1$ at $x = a$. For any given partition of $a \leqslant x \leqslant b$, say $x_1, \ldots, x_{m+1}$, we can seek to approximate the solution $\mathbf{y} = \mathbf{f}(x)$ of the system (11–4), subject to these initial conditions, by replacing this equation by the set of algebraic equations (or difference equations)

$$\mathbf{y}_{j+1} = \mathbf{y}_j + \mathbf{G}(x_j, \mathbf{y}_j)h_j \qquad \text{for } j = 1, 2, \ldots, m \tag{11-5}$$

It can be seen that, starting with the value $\mathbf{y}_1$, equation (11–5) can be used to calculate $\mathbf{y}_j$ successively at the points $j = 2, 3, \ldots, m+1$. We hope that the vectors $\mathbf{y}_j$ will provide good approximations to the values $\mathbf{f}(x_j)$ of the solution to equation (11–4) at the points x_j.

The discretization error which occurs when using a difference equation to integrate a differential equation across a single step, $T(x_j, h_j)$, is called the *local truncation error* or the *truncation error per step*. Thus, the truncation error per step is the error induced by using equation (11–5) to calculate $\mathbf{f}(x_j + h_j)$ approximately from the value $\mathbf{f}(x_j) = \mathbf{y}_j$ or, symbolically,

$$T(x_j, h_j) = |\, \mathbf{y}_{j+1} - \mathbf{f}(x_j + h_j)\,|$$

Hence, upon substituting in equation (11–5) with $\mathbf{y}_j = \mathbf{f}(x_j)$, we get

$$T(x_j, h_j) = |\, \mathbf{f}(x_j) - \mathbf{f}(x_j + h_j) + \mathbf{G}(x_j, \mathbf{f}(x_j))h_j\,|$$

But Taylor's theorem shows that [111]

$$\mathbf{f}(x_j + h_j) = \mathbf{f}(x_j) + \left(\frac{d\mathbf{f}}{dx} \right)_{x_j} h_j + O(h_j^2)$$

Hence,

$$T(x_j, h_j) = \left| \mathbf{G}(x_j, \mathbf{f}(x_j)) - \left(\frac{d\mathbf{f}}{dx} \right)_{x_j} \right| h_j + O(h_j^2)$$

But since, by hypothesis, $\mathbf{f}(x)$ satisfies the differential equation (11–4), the first term vanishes and we obtain

$$T(x_j, h_j) = O(h_j^2) \qquad \text{as } h_j \to 0 \tag{11–6}$$

More generally, we can attempt to approximate the solution to equation (11–4) by means of a difference equation of the form

$$\mathbf{y}_{j+1} = \mathbf{y}_j + \boldsymbol{\Phi}(x_j, \mathbf{y}_j; h_j)h_j \qquad \text{for } j = 1, 2, \ldots, m \tag{11–7}$$

where the function $\boldsymbol{\Phi}$ is to be chosen so that in some sense the solutions to equation (11–7) provide good approximations to the solutions of equation (11–4) for sufficiently small step size h_j. In order that this be the case, we certainly must require that

$$\boldsymbol{\Phi}(x, \mathbf{y}; h) \to \mathbf{G}(x, \mathbf{y}) \qquad \text{as } h \to 0 \tag{11–8}$$

If we again let $\mathbf{y} = \mathbf{f}(x)$ be a solution to equation (11–4), the truncation error per step $T(x_j, h_j)$ is

---

[111] The concept of order and the symbol O are introduced in chapter 9.

$$T(x_j, h_j) = |\mathbf{f}(x_j) - \mathbf{f}(x_j + h_j) + \mathbf{\Phi}(x_j, \mathbf{f}(x_j); h_j)h_j|$$

The same argument as was used in obtaining equation (11–6) used together with condition (11–8) now implies that

$$T(x_j; h_j) = o(h_j) \qquad \text{as } h_j \to 0 \tag{11–9}$$

This shows that the truncation error per step goes to zero faster than the mesh size h_j. Other things being equal, it is of course desirable to have the truncation error per step approach zero at the fastest possible rate as $h_j \to 0$. In order to have some measure of this rate we define the *order of a* given *difference scheme* to be the largest number p such that

$$T(x_j; h_j) = O(h_j^{p+1}) \qquad \text{as } h_j \to 0$$

Equation (11–6) shows that the order of the difference scheme (11–5) is 1. And, more generally, equation (11–9) shows that the order of any difference scheme of the type (11–7) which satisfies condition (11–8) is greater than zero. The numerical method corresponding to the difference scheme (11–5) is called *Euler's method*. Although this method is very simple, it is prone to round-off errors and is therefore infrequently used.

The truncation error per step can be used to obtain a bound on the (cumulative) truncation error for difference equations such as equation (11–7), that is, difference equations which determine the solution at the point x_{j+1} only in terms of quantities from the preceding step. Instead of considering the general system (11–4), we consider, for simplicity, only the single differential equation

$$\frac{dy}{dx} = G(x, y) \tag{11–10}$$

Suppose $G(x, y)$ satisfies the Lipschitz condition with respect to y

$$|G(x, y) - G(x, \zeta)| \leqslant M|y - \zeta|$$

and let

$$\Delta \equiv \max \frac{T(x_j, h_j)}{h_j} \qquad \text{for } j = 1, 2, \ldots, m$$

Then it is shown in reference 4 (p. 181) that the truncation error E_j is, at most,

$$\frac{\Delta}{M} \left(e^{|x_j - x_1|M} - 1 \right)$$

that is,

$$E_j \leq \frac{\Delta}{M} \left(e^{|x_j - x_1|M} - 1 \right)$$

Now for any difference scheme of order p, there exist constants D_j independent of the mesh size h_j such that

$$\frac{T(x_j, h_j)}{h_j} \leq D_j h_j^p$$

And if we put

$$\left.\begin{aligned} \epsilon &= \max h_j \\[2mm] D &\equiv \max D_j \end{aligned}\right\} \qquad \text{for } j = 1, 2 \ldots, m$$

then $\Delta \leq D\epsilon^p$ and the truncation error satisfies the inequality

$$E_j \leq \frac{D\epsilon^p}{M} \left(e^{|x_j - x_1|M} - 1 \right) \qquad \text{for } j = 1, 2, \ldots, m$$

which shows that the (cumulative) truncation error is of order ϵ^p when the difference scheme is of order p.

In principle, it is easy to derive formulas for numerical integration of the system of equations (11–4) which are of an arbitrarily high order. This can be accomplished, for example, by the *method of Taylor's series*. Instead of applying this method to the general system of first-order equations (11–4), we shall again, for simplicity, consider only the single first-order equation (11–10). There is no difficulty in extending the ideas to the general system (11–4).

Let $y=f(x)$ be a solution to equation (11–10) and let $g(x, y)$ be any r-times continuously differentiable function of x and y. Taking the total derivative of g with respect to x along the curve which is obtained by plotting $y=f(x)$ gives

$$g_1 \equiv \frac{dg}{dx} = \frac{\partial g}{\partial x} + \frac{\partial g}{\partial y}\frac{dy}{dx} = \frac{\partial g}{\partial x} + G\frac{\partial g}{\partial y}$$

Applying this formula to the function $g_1(x, y)$ gives

$$g_2 \equiv \frac{d^2 g}{dx^2} = \frac{dg_1}{dx} = \frac{\partial g_1}{\partial x} + G\frac{\partial g_1}{\partial y} = \left(\frac{\partial}{\partial x} + G\frac{\partial}{\partial y}\right)\left(\frac{\partial}{\partial x} + G\frac{\partial}{\partial y}\right)g \equiv \left(\frac{\partial}{\partial x} + G\frac{\partial}{\partial y}\right)^2 g$$

and, in general, we obtain

$$g_n \equiv \frac{d^n g}{dx^n} = \frac{dg_{n-1}}{dx} = \frac{\partial g_{n-1}}{\partial x} + G\frac{\partial g_{n-1}}{\partial y} = \left(\frac{\partial}{\partial x} + G\frac{\partial}{\partial y}\right)^n g \qquad \text{for } n = 1, 2, \ldots, r$$

Thus, in the special case when $g(x, y) = G(x, y)$, this equation becomes, in view of the differential equation (11–10),

$$\frac{d^2 y}{dx^2} = G_x + GG_y \tag{11–11}$$

$$\frac{d^3 y}{dx^3} = G_{xx} + 2GG_{xy} + G^2 G_{yy} + G_x G_y + GG_y^2 \tag{11–12}$$

.

.

.

$$\frac{d^{n+1}y}{dx^{n+1}} = \left(\frac{\partial}{\partial x} + G \frac{\partial}{\partial y} \right)^n G \qquad (11\text{--}13)$$

$$\cdot$$
$$\cdot$$
$$\cdot$$

$$\frac{d^{r+1}y}{dx^{r+1}} = \left(\frac{\partial}{\partial x} + G \frac{\partial}{\partial x} \right)^r G$$

Now suppose that G is $p-1$ times continuously differentiable and put

$$\Phi(x, y; h) \equiv \sum_{n=0}^{p-1} \frac{h^n}{(n+1)!} \left(\frac{\partial}{\partial x} + G \frac{\partial}{\partial y} \right)^n G \qquad (11\text{--}14)$$

Then, at least in principle, Φ can be calculated for any integer p simply by differentiating the given function G. Hence, when $y=f(x)$ is a solution to equation (11–10), it follows from equation (11–13) that

$$\Phi(x, f(x); h) = \sum_{n=0}^{p-1} \frac{h^n}{(n+1)!} \frac{d^{n+1}y}{dx^{n+1}} \qquad (11\text{--}15)$$

Thus, the truncation error per step, which is incurred when the difference equation

$$y_{j+1} = y_j + \Phi(x_j, y_j; h_j) h_j \qquad \text{for } j=1, 2, \ldots, m \qquad (11\text{--}16)$$

is solved to obtain an approximation to the exact solution, $y=f(x)$, of the differential equation (11–10), is

$$T(x_j, h_j) = | f(x_j) - f(x_j + h_j) + \Phi(x_j, f(x_j); h_j) h_j |$$

But inserting equation (11–15) shows that

$$T(x_j, h_j) = \left| f(x_j) - f(x_j + h_j) + \sum_{n=0}^{p-1} \frac{h_j^{n+1}}{(n+1)!} \left(\frac{d^{n+1}y}{dx^{n+1}} \right)_{x_j} \right|$$

Hence, upon applying Taylor's theorem we find that

$$T(x_j, h_j) = O(h_j^{p+1})$$

which shows that the formula (11–16) with Φ determined by equation (11–14) is of order p. The method of Taylor's series is efficient for linear systems or even for equations where G is a polynomial of low degree in x and y. However, as can be seen from equations (11–11) and (11–12), the method usually becomes extremely complex. In order to avoid this complication due to the successive differentiation and at the same time to preserve the increased accuracy which is afforded by using the Taylor's series method, a technique introduced by Runge, Kutta, and Heun known as the *Runge-Kutta method* can be employed.

In this case, the function Φ in equation (11–7) is taken to be of the form

$$\Phi(x_j, y_j; h_j) = \sum_{s=1}^{r} \alpha_s \mathbf{k}_s \tag{11–17}$$

where

$$\mathbf{k}_1 \equiv \mathbf{G}(x_j, \mathbf{y}_j) \tag{11–18}$$

and

$$\mathbf{k}_s \equiv \mathbf{G}\left(x_j + \mu_s h_j, \mathbf{y}_j + h_j \sum_{n=1}^{s-1} \lambda_{s-1, n} \mathbf{k}_n\right) \quad \text{for } s = 2, 3, \ldots, r \tag{11–19}$$

For any given integer r, the parameters α_s, μ_s, and $\lambda_{s, n}$ are to be determined in such a way that the order p of equation (11–7) is as large as possible.

For simplicity, we shall again restrict our attention to the single first-order equation

$$\frac{dy}{dx} = G(x, y) \qquad a \leqslant x \leqslant b \tag{11–20}$$

In this case, equations (11–7) and (11–17) to (11–19) reduce to

$$y_{j+1} = y_j + \Phi(x_j, y_j; h_j) h_j \qquad \text{for } j = 1, 2, \ldots, m \tag{11–21}$$

$$\Phi = \sum_{s=1}^{r} \alpha_s k_s \tag{11-22}$$

$$k_1 = G(x_j, y_j) \tag{11-23}$$

$$k_s = G\left(x_j + \mu_s h_j,\ y_j + h_j \sum_{n=1}^{s-1} \lambda_{s-1,n} k_n\right) \quad \text{for } s = 2, 3, \ldots, r \tag{11-24}$$

And if $y = f(x)$ is a solution to equation (11–20), the expression for $T(x_j, h_j)$, the truncation error per step, reduces to

$$T(x_j, h_j) = |f(x_j) - f(x_j + h_j) + \Phi(x_j, f(x_j); h_j) h_j| \tag{11-25}$$

Applying Taylor's theorem to the function $T(x_j, h_j)$ gives

$$T(x_j, h_j) = \sum_{m=0}^{p} \frac{1}{m!} \left[\frac{\partial^m T(x_j, h)}{\partial h^m}\right]_{h=0} h_j^m + O(h_j^{p+1})$$

But differentiating equation (11–25) m times gives

$$\left[\frac{\partial^m T(x_j, h)}{\partial h^m}\right]_{h=0} = \left| m\left(\frac{\partial^{m-1}\Phi}{\partial h_j^{m-1}}\right)_{h_j=0} - \left(\frac{d^m y}{dx^m}\right)_{x=x_j}\right|$$

Hence, the method will be of order p, provided that

$$m\left(\frac{\partial^{m-1}\Phi}{\partial h_j^{m-1}}\right)_{h_j=0} - \left(\frac{d^m y}{dx^m}\right)_{x=x_j} \begin{cases} \equiv 0 & \text{for } m = 1, 2, \ldots, p \\ \not\equiv 0 & \text{for } m = p+1 \end{cases} \tag{11-26}$$

If we substitute equation (11–13) and equations (11–22) to (11–24) into equation (11–26) and equate to zero the coefficients of all the independent partial derivatives of G, we will get a set of nonlinear equations for α_s, μ_s, and $\lambda_{s,n}$. For any given value of r, there will be a largest value of p for which these equations can be solved. For $1 \leq r \leq 4$, this value of p turns out to be equal to r.

There is a certain arbitrariness in the solutions of these equations for the constants α_s, μ_s, and $\lambda_{s,n}$. Thus, for $r = 2$, one of these constants can be chosen arbitrarily; and for $r = 3$ and $r = 4$, two of the constants can be chosen arbitrarily. The computations are carried out for the general case in reference 41.

DIFFERENTIAL EQUATIONS

Here we shall merely illustrate the method by considering the case where $r=3$. Then equations (11–22) to (11–24) reduce to

$$\Phi = \alpha_1 k_1 + \alpha_2 k_2 + \alpha_3 k_3 \qquad (11\text{–}27)$$

$$\left.\begin{aligned}
k_1 &= G(x_j, y_j) \\
k_2 &= G(x_j + \mu_2 h_j, y_j + h_j \lambda_{1,1} k_1) \\
k_3 &= G(x_j + \mu_3 h_j, y_j + h_j \lambda_{2,1} k_1 + h_j \lambda_{2,2} k_2)
\end{aligned}\right\} \qquad (11\text{–}28)$$

Expanding k_1, k_2, and k_3 in a Taylor series about $h_j = 0$ and retaining terms only up to h_j^2 give

$$\left.\begin{aligned}
k_1 &= G\,(x_j, y_j) \\[2ex]
k_2 &= G\,(x_j, y_j) + h_j\,(\mu_2 G_x + \lambda_{1,1} GG_y)_{\substack{x=x_j \\ y=y_j}} \\[2ex]
&\quad + h_j^2 \left(\frac{1}{2}\mu_2^2 G_{xx} + \mu_2 \lambda_{1,1} GG_{xy} + \frac{1}{2}\lambda_{1,1}^2 G^2 G_{yy}\right)_{\substack{x=x_j \\ y=y_j}} + O(h_j^3) \\[2ex]
k_3 &= G\,(x_j, y_j) + h_j\,(\mu_3 G_x + \lambda_{2,1} GG_y + \lambda_{2,2} GG_y)_{\substack{x=x_j \\ y=y_j}} \\[2ex]
&\quad + h_j^2 \left[\frac{1}{2}\mu_3^2 G_{xx} + \mu_3\,(\lambda_{2,1}+\lambda_{2,2})GG_{xy} + \frac{1}{2}(\lambda_{2,1}+\lambda_{2,2})^2 G^2 G_{yy}\right. \\[2ex]
&\qquad \left. + \lambda_{2,2}\,(\mu_2 G_x + \lambda_{1,1}GG_y)\,G_y\right]_{\substack{x=x_j \\ y=y_j}} + O(h_j^3)
\end{aligned}\right\} \quad (11\text{–}29)$$

Now it is clear that

$$\frac{1}{m!}\left(\frac{\partial^m \Phi}{\partial h_j^m}\right)_{h_j=0}$$

is the coefficient of $(h_j)^m$ in the Taylor series expansion of Φ about $h_j=0$. Hence, when equations (11-29) are substituted into equation (11-27), the functions

$$\frac{1}{m!}\left(\frac{\partial^m \Phi}{\partial h_j^m}\right)_{h_j=0}$$

for $m=0, 1, 2$ are simply the coefficients of h_j^0, h_j, and h_j^2, respectively, in the resulting expression. Thus,

$$\Phi(x_j, y_j; 0) = (\alpha_1 + \alpha_2 + \alpha_3) \, G(x_j, y_j) \tag{11-30}$$

$$\left(\frac{\partial \Phi}{\partial h_j}\right)_{h_j=0} = \alpha_2 \, (\mu_2 G_x + \lambda_{1,1} G G_y)_{\substack{x=x_j \\ y=y_j}} + \alpha_3 \, [\mu_3 G_x + (\lambda_{2,1} + \lambda_{2,2}) G G_y]_{\substack{x=x_j \\ y=y_j}} \tag{11-31}$$

$$\frac{1}{2}\left(\frac{\partial^2 \Phi}{\partial h_j^2}\right)_{h_j=0} = \alpha_2 \left(\frac{1}{2} \mu_3^2 G_{xx} + \mu_2 \lambda_{1,1} G G_{xy} + \frac{1}{2} \lambda_{1,1}^2 G^2 G_{yy} \right)_{\substack{x=x_j \\ y=y_j}}$$

$$+ \alpha_3 \left[\frac{1}{2} \mu_3^2 G_{xx} + \mu_3 (\lambda_{2,1} + \lambda_{2,2}) G G_{xy} + \frac{1}{2} (\lambda_{2,1} + \lambda_{2,2})^2 G^2 G_{yy} \right.$$

$$\left. + \lambda_{2,2} (\mu_2 G_x + \lambda_{1,1} G G_y) G_y \right]_{\substack{x=x_j \\ y=y_j}} \tag{11-32}$$

On the other hand, equations (11-26) show that

$$\Phi\left(x_j, y_j; 0\right) = \left(\frac{dy}{dx}\right)_{x=x_j}$$

$$2\left(\frac{\partial\Phi}{\partial h_j}\right)_{h_j=0} = \left(\frac{d^2y}{dx^2}\right)_{x=x_j}$$

$$3\left(\frac{\partial^2\Phi}{\partial h_j^2}\right)_{h_j=0} = \left(\frac{d^3y}{dx^3}\right)_{x=x_j}$$

Upon substituting equations (11–11), (11–12), (11–20), and (11–30) to (11–32) into these relations, we find that the resulting equations are satisfied identically in x_j and y_j only if the constants α_s, μ_s, and $\lambda_{s,n}$ are chosen so that the coefficients of all the independent derivatives of G vanish. This will occur if, and only if, the constants satisfy the following algebraic equations:

$$\alpha_1 + \alpha_2 + \alpha_3 = 1 \qquad\qquad \mu_2\alpha_2 + \mu_3\alpha_3 = \frac{1}{2}$$

$$\lambda_{1,1}\alpha_2 + (\lambda_{2,1} + \lambda_{2,2})\alpha_3 = \frac{1}{2} \qquad\qquad \mu_2^2\alpha_2 + \mu_3^2\alpha_3 = \frac{1}{3}$$

$$\mu_2\lambda_{1,1}\alpha_2 + \mu_3(\lambda_{2,1} + \lambda_{2,2})\alpha_3 = \frac{1}{3} \qquad\qquad \lambda_{1,1}^2\alpha_2 + (\lambda_{2,1} + \lambda_{2,2})^2\alpha_3 = \frac{1}{3}$$

$$\mu_2\lambda_{2,2}\alpha_3 = \frac{1}{6} \qquad\qquad \lambda_{1,1}\lambda_{2,2}\alpha_3 = \frac{1}{6}$$

These equations imply that

$$\lambda_{1,1} = \mu_2 \qquad\qquad \lambda_{2,1} + \lambda_{2,2} = \mu_3$$

and there are four independent equations which must be satisfied by the remaining six unknown constants. Hence, two of these constants can be chosen arbitrarily.

The most frequently used Runge-Kutta method is of the fourth order. The values of the constants α_s, μ_s, and $\lambda_{s,n}$ for the general vector equations

Table 11–1.—Choice of Parameters for Fourth-Order Runge-Kutta Method

Parameters	Standard Runge-Kutta method	Kutta's method
α_1	1/6	1/8
α_2	1/3	3/8
α_3	1/3	3/8
α_4	1/6	1/8
μ_2	1/2	1/3
μ_3	1/2	2/3
μ_4	1	1
$\lambda_{1,1}$	1/2	1/3
$\lambda_{2,1}$	0	−1/3
$\lambda_{2,2}$	1/2	1
$\lambda_{3,1}$	0	1
$\lambda_{3,2}$	0	−1
$\lambda_{3,3}$	1	1

(11–17) to (11–19) for the fourth-order Runge-Kutta method are listed in table 11–1 for two choices of the arbitrary constants, and because of its importance the complete formulas for the standard Runge-Kutta method are also listed:

$$y_{j+1} = y_j + \frac{h_j}{6}(k_1 + 2k_2 + 2k_3 + k_4)$$

$$k_1 = G(x_j, y_j)$$

$$k_2 = G\left(x_j + \frac{1}{2}h_j, y_j + \frac{1}{2}h_j k_1\right)$$

$$k_3 = G\left(x_j + \frac{1}{2}h_j, y_j + \frac{1}{2}h_j k_2\right)$$

$$k_4 = G(x_j + h_j, y_j + h_j k_3)$$

There appears to be only a slight advantage which can be gained by changing the choice of the arbitrary parameters.

Although the Runge-Kutta method involves fairly simple formulas, it has certain disadvantages. Thus, (1) the method is limited to the fourth or fifth order; (2) if the function $\mathbf{G}$ is complicated, the evaluation of the $\mathbf{k}_s$ for $s = 1, 2, 3, 4$ at each mesh point can be quite time consuming; (3) it will calculate a solution even at points of discontinuity without giving any indication that this has been done; and (4) there is no readily obtainable error analysis.

The lack of any error analysis for the fourth-order Runge-Kutta method can be partially compensated for by using certain rules of thumb. Thus, for example (see ref. 41), if the quantity

$$\left| \frac{\mathbf{k}_2 - \mathbf{k}_3}{\mathbf{k}_1 - \mathbf{k}_2} \right|$$

becomes much larger than a few hundredths at any point x_j, the step size h_j should be decreased.

11.2.1.2   *Multistep processes: Finite differences.* — Up to this point we have been discussing one-step difference equations, that is equations which determine the value of the dependent variable at the step x_{j+1} completely in terms of its value at the preceding step x_j. There are other types of difference equations which can be used, called *n-step* equations,[112] which utilize the values of the dependent variable at the first n preceding steps, say $x_j, x_{j-1}, \ldots, x_{j-n+1}$, to determine its value at the step x_{j+1}.

Before discussing these difference equations it is first convenient to introduce the concept of difference operator. Thus, the difference operators $\Delta, \nabla,$ and δ corresponding, respectively, to the *forward difference*, the *backward difference*, and the *central difference* are defined by

$$\Delta y(x) \equiv y(x+h) - y(x)$$

$$\nabla y(x) \equiv y(x) - y(x-h)$$

$$\delta y(x) \equiv y\left(x + \frac{h}{2}\right) - y\left(x - \frac{h}{2}\right)$$

If the function $y(x)$ is defined only on a finite set of points, say $x_1 < x_2 < \ldots < x_{m+1}$ with $x_{j+1} - x_j = h$ for $j = 1, 2, \ldots, m$, we shall sometimes write

---

[112] The associated numerical procedure is referred to as an *n-step* process.

$$y_j \equiv y(x_j) = y(x_1 + (j-1)h)$$

Then the notation for the first two difference operators becomes

$$\Delta y_j = y_{j+1} - y_j \qquad \nabla y_j = y_j - y_{j-1}$$

These operators arise in approximating the derivative of functions. The nature of this approximation can be seen from the relations

$$\lim_{h \to 0} \frac{\Delta y(x)}{h} = \lim_{h \to 0} \frac{\nabla y(x)}{h} = \lim_{h \to 0} \frac{\delta y(x)}{2h} = \frac{dy(x)}{dx}$$

Applying these operators twice in succession gives the second differences

$$\Delta^2 y(x) = \Delta[\Delta y(x)] = \Delta[y(x+h) - y(x)] = y(x+2h) - 2y(x+h) + y(x)$$

$$\nabla^2 y(x) = y(x) - 2y(x-h) + h(x-2h)$$

$$\delta^2 y(x) = y(x+h) - 2y(x) + y(x-h)$$

These formulas can be used to provide approximations to the second derivative since

$$\frac{d^2 y(x)}{dx^2} = \lim_{h \to 0} \frac{\Delta^2 y(x)}{h^2} = \lim_{h \to 0} \frac{\nabla^2 y(x)}{h^2} = \lim_{h \to 0} \frac{\delta^2 y(x)}{h^2}$$

There are various manipulations that can be performed with these and other difference operators which are sometimes useful for obtaining finite difference equations from differential equations. A fairly detailed discussion of this is given in Hildebrand (ref. 44).

Now consider the general nth-order differential equation in normal form

$$\frac{d^n y}{dx^n} = G\left(x, y, \frac{dy}{dx}, \ldots, \frac{d^{n-1}y}{dx^{n-1}}\right) \qquad a \leqslant x \leqslant b \tag{11-33}$$

Suppose that the interval $a \leq x \leq b$ has a partition $x_1, x_2, \ldots, x_{m+1}$ for which all the subintervals $x_j \leq x \leq x_{j+1}$ have the same length h. Upon replacing each derivative $(d^k y)/(dx^k)$ for $k = 1, 2, \ldots, n$ in equation (11–33) by its forward difference approximation $h^{-k} \Delta^k y_j$, where y_j is the approximation to $y[a + (j-1)h] = y(x_j)$, we obtain the difference equation

$$\Delta^n y_j = h^n G\left(x_j, y_j, \frac{1}{h} \Delta y_j, \ldots, \frac{1}{h^{n-1}} \Delta^{n-1} y_j\right) \qquad (11\text{–}34)$$

However, since $\Delta^k y_j$ is a linear combination of $y_j, y_{j+1}, \ldots, y_{j+k}$, this equation is essentially of the form

$$y_{j+m} = \phi(j, y_j, y_{j+1}, \ldots, y_{j+m-1}; h)$$

which is clearly an n-step difference equation.

In fact, even first-order normal differential equations can lead to n-step difference equations. Thus, consider the differential equation

$$\frac{dy}{dx} = G(x, y)$$

and approximate the derivative by the central difference $(y_{j+1} - y_{j-1})/2h$ to obtain the difference equation

$$y_{j+1} = y_{j-1} + 2hG(x_j, y_j)$$

which is clearly a two-step equation.

Notice that, in order to start the solution of an n-step method, we must have a prior knowledge of the values of $y_1, y_2, y_3, \ldots, y_n$, that is, the values (or approximate values) of the solution at the first n mesh points. A one-step method requires only a knowledge of the initial value y_1. Hence, when $n > 1$ it is in many instances necessary to introduce some auxiliary method for determining these values. The one-step methods are, therefore, said to be *self-starting*. In a one-step method, the mesh size $h_j \equiv x_{j+1} - x_j$ can be varied at each step. With a multistep method, it must usually remain fixed.

11.2.2 Implicit Methods: Predictor-Corrector

In the methods discussed so far, the value of the dependent variable y_{j+1} at the step x_{j+1} is determined *explicitly* in terms of its values at one or more preceding steps. We can, therefore, calculate the values of the dependent variable recursively. Hence, such methods are called *explicit methods*. There are other methods, however, in which the formulas for calculating y_{j+1} from the values of the dependent variable at the preceding steps are not solved explicitly for y_{j+1} but determine this variable only implicitly. Such methods are, therefore, called *implicit methods*.

In order to see how difference equations of this type arise, let

$$a = x_1 < x_2 < \ldots < x_{m+1} = b$$

be a partition of the interval $a \leqslant x \leqslant b$; suppose that each subinterval $x_j \leqslant x \leqslant x_{j+1}$ has the same length, say h; and consider the integral $\int_a^b f(x)\,dx$. Recall that this integral can be evaluated numerically in an approximate fashion by using either the trapezoidal rule

$$\int_a^b f(x)\,dx \approx \sum_{j=1}^{m} \frac{1}{2}\,[f(x_{j+1}) + f(x_j)]h$$

or by using Simpson's rule

$$\int_a^b f(x)\,dx \approx \sum_{j=1}^{m-1} \frac{1}{3}\,[f(x_{j+2}) + 4f(x_{j+1}) + f(x_j)]h$$

Now, consider the first-order differential equation

$$\frac{dy}{dx} = G(x, y) \tag{11–35}$$

Integrating both sides of this equation first between x_j and x_{j+1} and then between x_j and x_{j+2} gives, respectively,

$$y_{j+1} = y_j + \int_{x_j}^{x_{j+1}} G(x, y(x))\,dx \tag{11-36}$$

and

$$y_{j+2} = y_j + \int_{x_j}^{x_{j+2}} G(x, y(x))\,dx \tag{11-37}$$

Upon using the trapezoidal rule for evaluating the integral in equation (11–36) and Simpson's rule for evaluating the integral in equation (11–37), we get the following finite difference equations for approximating the solution to equation (11–35):

$$y_{j+1} = y_j + [G(x_{j+1}, y_{j+1}) + G(x_j, y_j)]\frac{h}{2} \tag{11-38}$$

and

$$y_{j+2} = y_j + [G(x_{j+2}, y_{j+2}) + 4G(x_{j+1}, y_{j+1}) + G(x_j, y_j)]\frac{h}{3} \tag{11-39}$$

Notice that in the first of these equations, y_{j+1} appears not only explicitly but is involved implicitly through G on the right side. In general, it will not be possible to solve this equation to obtain an explicit formula for y_{j+1}. Similar remarks, of course, apply to the second equation. If, instead of treating the single first-order normal equation (11–35), we consider the general normal system

$$\frac{d\mathbf{y}}{dx} = \mathbf{G}(x, \mathbf{y}) \tag{11-40}$$

the same arguments would lead to the difference equations

$$\mathbf{y}_{j+1} = \mathbf{y}_j + [\mathbf{G}(x_{j+1}, \mathbf{y}_{j+1}) + \mathbf{G}(x_j, \mathbf{y}_j)]\frac{h}{2} \tag{11-41}$$

and

$$\mathbf{y}_{j+2} = \mathbf{y}_j + [\mathbf{G}(x_{j+2}, \mathbf{y}_{j+2}) + 4\mathbf{G}(x_{j+1}, \mathbf{y}_{j+1}) + \mathbf{G}(x_j, \mathbf{y}_j)] \frac{h}{3} \quad (11\text{--}42)$$

Notice that, in equation (11–41), $\mathbf{y}_{j+1}$ is determined completely in terms of its value at the preceding step; whereas, in equation (11–42), $\mathbf{y}_{j+2}$ is determined in terms of the values of the dependent variable at the preceding two steps. Since these formulas cannot usually be solved explicitly for the dependent variable, it is usually necessary to resort to an iteration process at each step to find this variable. Thus, let

$$\mathbf{U} \ (\mathbf{y}_{j+1}) \equiv \mathbf{y}_j + [\mathbf{G} \ (x_{j+1}, \mathbf{y}_{j+1}) + \mathbf{G} \ (x_j, \mathbf{y}_j)] \frac{h}{2}$$

Then equation (11–41) can be written as

$$\mathbf{y}_{j+1} = \mathbf{U} \ (\mathbf{y}_{j+1}) \quad (11\text{--}43)$$

Suppose that by some means a fairly good guess at the solution $\mathbf{y}_{j+1}$ of equation (11–41) can be made, say $\mathbf{y}_{j+1}^{(0)}$. Substituting this into equation (11–43) gives a better approximation $\mathbf{y}_{j+1}^{(1)}$ to the solution, given by $\mathbf{y}_{j+1}^{(1)} = \mathbf{U} \ (\mathbf{y}_{j+1}^{(0)})$.

Proceeding in this manner, we obtain the sequence of approximations

$$\mathbf{y}_{j+1}^{(1)} = \mathbf{U} \ (\mathbf{y}_{j+1}^{(0)}), \quad \mathbf{y}_{j+1}^{(2)} = \mathbf{U} \ (\mathbf{y}_{j+1}^{(1)}), \quad \mathbf{y}_{j+1}^{(3)} = \mathbf{U} \ (\mathbf{y}_{j+1}^{(2)}), \ \ldots$$

which we hope will converge fairly rapidly to the solution of equation (11–41). A good choice of the initial approximation $\mathbf{y}_{j+1}^{(0)}$ can be obtained by using Euler's formula, equation (11–5), to obtain

$$\mathbf{y}_{j+1}^{(0)} = \mathbf{y}_j + \mathbf{G} \ (x_j, \mathbf{y}_j) \ h \quad (11\text{--}44)$$

In practice, instead of solving equation (11–41) accurately for $\mathbf{y}_{j+1}$ by performing many iterations, we can obtain the same accuracy with much less work by taking a finer mesh size h and performing only one or two iterations. If only one iteration is performed, the method is called the *improved Euler method*. In this case, we first compute the vector $\mathbf{Y}_{j+1}$, called the *predictor*, from the formula

$$\mathbf{Y}_{j+1} = \mathbf{y}_j + \mathbf{G}\,(x_j, \mathbf{y}_j)\,h \qquad\qquad (11\text{--}45)$$

called the *predictor formula*, and then substitute it into the formula

$$\mathbf{y}_{j+1} = \mathbf{y}_j + [\mathbf{G}\,(x_{j+1}, \mathbf{Y}_{j+1}) + \mathbf{G}\,(x_j, \mathbf{y}_j)]\,\frac{h}{2} \qquad\qquad (11\text{--}46)$$

called the *corrector formula*, to determine $\mathbf{y}_{j+1}$. Thus, in effect, $\mathbf{y}_{j+1}$ is calculated from $\mathbf{y}_j$ in two steps instead of one. Of course, we can apply the same procedure to equation (11–42). This leads to *Milne's method*.

These two methods are examples of the *predictor-corrector methods*. Other predictor-corrector methods differ from these only with respect to the polynomial interpolation formulas from which the predictor and corrector formulas are derived. A commonly used predictor formula is the *Adams-Bashforth* formula (see ref. 45)

$$\mathbf{Y}_{j+1} = \mathbf{y}_j + \frac{h}{24}\,(55\mathbf{G}_j - 59\mathbf{G}_{j-1} + 37\mathbf{G}_{j-2} - 9\mathbf{G}_{j-3})$$

where we have put $\mathbf{G}_j \equiv \mathbf{G}(x_j, \mathbf{y}_j)$. This formula is most frequently used in conjunction with the *Adams-Moulton* corrector formula (see ref. 45)

$$\mathbf{y}_{j+1} = \mathbf{y}_j + \frac{h}{24}\,[9\mathbf{G}(x_{j+1}, \mathbf{Y}_{j+1}) + 19\mathbf{G}_j - 5\mathbf{G}_{j-1} + \mathbf{G}_{j-2}]$$

These formulas have a higher order of accuracy than the Euler's formulas. However, they are not self-starting. Also, unlike the Runge-Kutta method they cannot be easily used alone with a variable mesh size. These difficulties are frequently alleviated in practice by using the Runge-Kutta method to obtain the starting values and also to compute the solution for the first few mesh points after the step size has been changed. However, the predictor-corrector methods can, in the case of complicated equations, result in a considerable savings in computer time over the Runge-Kutta method. In addition, it is usually possible with predictor-corrector methods to monitor the error as the calculation proceeds.

Another difficulty with the predictor-corrector methods is that in some cases, they are subject to certain types of instabilities which do not occur when the Runge-Kutta method is used. This instability manifests itself first by resulting in an error which is larger than expected; and when one attempts to reduce this error by decreasing the step size, the error actually increases. A more detailed but elementary discussion of this instability is given in reference 45.

11.3 BOUNDARY-VALUE PROBLEMS

11.3.1 Linear Equations

11.3.1.1 *Use of superposition.*—The methods discussed previously have all been methods for solving initial-value problems. However, in the case of linear equations, these methods can be used in conjunction with the super-position principle to obtain solutions to boundary-value problems. Thus, in order to solve a boundary-value problem for a linear differential equation, we need only solve numerically the same number of initial-value problems as the order of the differential equation, provided these problems are chosen in such a way that their solutions are linearly independent of one another. It is easily seen from section 1.6 that this can always be done by choosing the initial conditions of these problems so that they have a nonzero Wronskian determinant. Then any boundary-value problem can be solved by forming a linear combination of these solutions with the constants adjusted numerically to satisfy the imposed boundary conditions. This is a particular example of how some a priori knowledge of the properties of the solutions of the equations to be solved can be utilized to simplify the numerical procedure for obtaining these solutions.

11.3.1.2 *Finite differences.*—Another method for solving linear boundary-value problems is the *method of finite differences*. In order to illustrate this method, let us consider the second-order linear equation

$$\frac{d^2y}{dx^2} + p(x)\,\frac{dy}{dx} + q(x)y = r(x) \qquad a \leqslant x \leqslant b \qquad (11\text{–}47)$$

subject to the boundary conditions

$$\left.\begin{aligned} y(a) &= A \\ y(b) &= B \end{aligned}\right\} \qquad (11\text{–}48)$$

DIFFERENTIAL EQUATIONS

Let $a = x_0 < x_1 < \ldots < x_{m+1} = b$ be a partition of $a \leq x \leq b$ with equal mesh size h.

Upon approximating the derivatives by appropriate central differences, we obtain the following difference-equation approximation to equation (11–47)

$$\frac{y_{j+1} - 2y_j + y_{j-1}}{2h} + \frac{y_{j+1} - y_{j-1}}{2h} p_j + y_j q_j = r_j \qquad \text{for } j = 1, 2, \ldots, m$$

where we have put $p_j \equiv p(x_j)$, $q_j \equiv q(x_j)$, and $r_j \equiv r(x_j)$. This can be written as

$$\left(1 - \frac{h}{2} p_j\right) y_{j-1} + \left(h^2 q_j - 2\right) y_j + \left(1 + \frac{h}{2} p_j\right) y_{j+1} = h^2 r_j \qquad \text{for } j = 1, 2, \ldots, m$$

Upon using the boundary conditions (11–48) to replace y_0 by A and y_{m+1} by B, we obtain the following set of m equations in the m unknowns $y_1, y_2, \ldots, y_m$:

$$\left(h^2 q_1 - 2\right) y_1 + \left(1 + \frac{h}{2} p_1\right) y_2 = h^2 r_1 + \left(\frac{h}{2} p_1 - 1\right) A$$

$$\left(1 - \frac{h}{2} p_2\right) y_1 + \left(h^2 q_1 - 2\right) y_2 + \left(1 + \frac{h}{2} p_2\right) y_3 = h^2 r_2$$

$$\cdot$$
$$\cdot$$
$$\cdot$$

$$\left(1 - \frac{h}{2} p_{m-1}\right) y_{m-2} + \left(h^2 q_{m-1} - 2\right) y_{m-1} + \left(1 + \frac{h}{2} p_{m-1}\right) y_m = h^2 r_{m-1}$$

$$\left(1 - \frac{h}{2} p_m\right) y_{m-1} + \left(h^2 q_1 - 2\right) y_m = h^2 r_m - \left(1 + \frac{h}{2} p_m\right) B$$

This equation can be written in matrix form as

$$\mathbf{M}\mathbf{y} = \mathbf{c} \qquad\qquad (11\text{--}49)$$

where $\mathbf{y}$ is the vector $\mathbf{y} = (y_1, y_2, \ldots, y_m)$, $\mathbf{M}$ is a matrix of the form

$$\mathbf{M} = \begin{bmatrix} \beta_1 & \gamma_1 & 0 & 0 & \dots & 0 & 0 & 0 & 0 \\ \alpha_2 & \beta_2 & \gamma_2 & 0 & \dots & 0 & 0 & 0 & 0 \\ 0 & \alpha_3 & \beta_3 & \gamma_3 & \dots & 0 & 0 & 0 & 0 \\ \cdot & \cdot & \cdot & \cdot & & \cdot & & \cdot & \cdot \\ \cdot & \cdot & \cdot & \cdot & & \cdot & & \cdot & \cdot \\ \cdot & \cdot & \cdot & \cdot & & \cdot & & \cdot & \cdot \\ 0 & 0 & 0 & 0 & \dots & 0 & \alpha_{m-1} & \beta_{m-1} & \gamma_{m-1} \\ 0 & 0 & 0 & 0 & \dots & 0 & 0 & \alpha_m & \beta_m \end{bmatrix}$$

and $\mathbf{c}$ is a vector. The entries in $\mathbf{M}$ and $\mathbf{c}$ are all known since p, q, and r are known functions. For obvious reasons, the matrix $\mathbf{M}$ is said to be *tridiagonal*. Tridiagonal matrices can be numerically inverted easily and quickly; and, therefore, equation (11–49) can readily be solved by inverting $\mathbf{M}$ to find the solution vector $\mathbf{y}$. A particularly convenient technique is known as the *line inversion* method. In setting up finite difference problems it is not always possible to obtain tridiagonal or even *n-diagonal* matrices. However, whenever possible, the difference equations should be set up to obtain n-diagonal matrices since they can usually be inverted more easily than other types of matrices. In addition, when the system is programed for a digital computer, it is not necessary to define all m^2 locations of the coefficient matrix $\mathbf{M}$; in the case of a tridiagonal matrix, for example, only $3m$ locations need be allocated to $\mathbf{M}$ while performing the inversion. In any case the matrices which arise in the finite difference methods may usually be inverted by either implicit or explicit means. For a discussion of the various methods for accomplishing this, the reader is referred to references 46 and 47. It sometimes happens, however, that it is not possible to invert these matrices; it is then necessary to use an iterative process to solve the matrix equation (see ref. 47).

11.3.2 Nonlinear Equations

11.3.2.1 *Shooting methods.*—Boundary-value problems for nonlinear equations can be solved by using "*shooting methods.*" To use these methods the problem is first transformed to an initial-value problem by guessing enough additional initial conditions at one boundary to allow the integration to proceed across the interval to the other boundary. In the initial trial the specified boundary conditions at the second boundary are unlikely to be met. But, by adjusting the additional initial conditions imposed at the first boundary, it is possible to come closer to satisfying the prescribed boundary conditions by carrying through a number of iterations.

The required adjustments to the additional initial conditions can be made in a number of different ways. Perhaps the most common of these is linear interpolation. In order to illustrate the ideas involved, consider a two-point boundary-value problem for a second-order system on the interval $a \leq x \leq b$ in which the boundary conditions $y(a) = A$ and $y(b) = B$ are specified. And suppose that the initial slope $y'(a)$ is to be adjusted until the second boundary condition is satisfied. When two values of the initial slope $y_1'(a)$ and $y_2'(a)$ have been found which lead to the two boundary values $y_1(b)$ and $y_2(b)$, respectively, such that

$$y_1(b) \leq B \leq y_2(b)$$

the next trial value of $y'(a)$ is determined by linear interpolation by using the prescription

$$\frac{y'(a) - y_1'(a)}{y_2'(a) - y_1'(a)} = \frac{B - y_1(b)}{y_2(b) - y_1(b)}$$

If this new value of $y'(a)$ leads to a value $y(b)$ which is either smaller or larger than B, it can be used together with either $y_2(b)$ or $y_1(b)$, respectively, to repeat the process. The process can be continued until the second boundary condition is satisfied to within the desired accuracy.

One difficulty with shooting methods which sometimes occurs is that the differential equation is so unstable that it "blows up" before the initial-value problem can be completely integrated. In such cases the process of quasi-linearization can be used.

11.3.2.2 *Quasi-linearization.*—Nonlinear boundary-value problems can also be solved by reducing them to linear problems by a process of quasi-linearization. This process consists of replacing the original equation by a sequence of linear equations in such a way that the sequence of solutions to these equations converges to the solution of the original equation. Thus, consider the first-order normal equation

$$\frac{dy}{dx} = G(x, y) \tag{11-50}$$

If $\tilde{y}$ is close to y, we might anticipate from Taylor's theorem that

$$G(x, y) \approx G(x, \tilde{y}) + \frac{\partial G}{\partial y}(x, \tilde{y})(y - \tilde{y})$$

We, therefore, replace equation (11–50) by the sequence

$$\frac{dy^{(n+1)}}{dx} = G(x, y^{(n)}) + \frac{\partial G}{\partial y}(x, y^{(n)})[y^{(n+1)} - y^{(n)}] \qquad \text{for } n = 0, 1, 2, \ldots$$

$$(11\text{–}51)$$

of linear equations for $y^{(n+1)}$. To use these equations we, first, choose a reasonable guess for $y^{(0)}$, calculate $y^{(1)}$, insert it in the right side, and calculate $y^{(2)}$. Proceeding in this manner, we obtain a sequence of solutions $y^{(n)}$ which we hope will converge to the solution y of the original problem. Since each equation (11–51) is linear, it can be handled by the methods described previously. These ideas are easily generalized to the first-order normal system

$$\frac{d\mathbf{y}}{dx} = \mathbf{G}(x, \mathbf{y})$$

and, therefore, to any normal system of differential equations. The basic reference on the quasi-linearization process is Bellman and Kalaba (ref. 48), in which are discussed various conditions which can be imposed on the functions $\mathbf{G}$ to ensure that the iterations converge.

Although all the most commonly used methods are described in this chapter, it is impossible in a single chapter to touch upon all the available techniques. For a more complete treatment of the subject, the reader is referred to the references cited.

REFERENCES

1. STAKGOLD, IVAR: Boundary Value Problems of Mathematical Physics. Vol. I. Macmillan Co., 1967.

2. FRIEDMAN, AVNER: Generalized Functions and Partial Differential Equations. Prentice-Hall, Inc., 1963.

3. INCE, EDWARD L.: Ordinary Differential Equations. Dover Publications, 1953.

4. BIRKHOFF, GARRETT; and ROTA, GIAN-CARLO: Ordinary Differential Equations. Ginn and Co., 1962.

5. GREENSPAN, DONALD: Theory and Solution of Ordinary Differential Equations. Macmillan Co., 1960.

6. KAPLAN, WILFRED: Advanced Calculus. Addison-Wesley Publ. Co., Inc., 1952.

7. KAPLAN, WILFRED: Ordinary Differential Equations. Addison-Wesley Publ. Co., Inc., 1958.

8. GOLDSTEIN, MARVIN E.; and ROSENBAUM, BURT M.: Introduction to Abstract Analysis. NASA SP-203, 1969.

9. MORLEY, F. V.: A Curve of Pursuit. Am. Math. Monthly, vol. 28, 1921, pp. 55-93.

10. DAVIS, HAROLD T.: Introduction to Nonlinear Differential and Integral Equations. Dover Publications, 1960.

11. PAINLEVÉ, P.: On Differential Equations of the Second and of Higher Order, the General Integral of Which is Uniform. Acta Math., vol. 25, 1902, pp. 1-85.

12. HERBST, ROBERT T.: The Equivalence of Linear and Nonlinear Differential Equations. Proc. Am. Math. Soc., vol. 7, 1956, pp. 95-97.

13. GERGEN, J. J.; and DRESSEL, F. G.: Second Order Linear and Nonlinear Differential Equations. Proc. Am. Math. Soc., vol. 16, 1965, pp. 767-773.

14. PINNEY, EDMUND: The Nonlinear Differential Equation $y'' + p(x)y + cy^{-3} = 0$. Proc. Am. Math. Soc., vol. 1, 1950, p. 681.

15. RAINVILLE, EARL D.: Intermediate Differential Equations. Second ed., Macmillan Co., 1964.

16. MURPHY, GEORGE M.: Ordinary Differential Equations and Their Solutions. D. Van Nostrand Co., Inc., 1960.

17. KAMKE, E.: Differentialgleichungen, Lösungsmethoden und Lösungen I Gewöhnliche Differentialgleichungen. Akademische Verlagsgesellschaft Becker & Erler kom.-ges., Leipzig, 1943.

18. CHURCHILL, RUEL V.: Complex Variables and Applications. Second ed., McGraw-Hill Book Co., Inc., 1960.

19. AHLFORS, LARS V.: Complex Analysis. McGraw-Hill Book Co., Inc., 1953.

20. CARRIER, GEORGE F.; KROOK, MAX; and PEARSON, CARL E.: Functions of a Complex Variable. McGraw-Hill Book Co., Inc., 1966.

21. NEHARI, ZEEV: Conformal Mapping. McGraw-Hill Book Co., Inc., 1962.

22. MORETTI, GINO: Functions of a Complex Variable. Prentice-Hall, Inc., 1964.

23. RAINVILLE, EARL D.: Infinite Series. Macmillan Co., 1967.

24. POOLE, E. G. C.: Introduction to the Theory of Linear Differential Equations. Dover Publications, 1936.

25. WHITTAKER, EDMUND T.; and WATSON, GEORGE N.: A Course in Modern Analysis. Fourth ed., Cambridge Univ. Press, 1927.

26. ERDÉLYI, A.; MAGNUS, W.; OBERHETTINGER, F.; and TRICOMI, F.: Higher Transcendental Functions. Vol. 2, McGraw-Hill Book Co., Inc., 1953.

27. RAINVILLE, EARL D.: Special Functions. Macmillan Co., 1960.

28. BUCHHOLZ, HERBERT: The Confluent Hypergeometric Function with Special Emphasis on Its Applications. Springer-Verlag, 1969.

29. WATSON, GEORGE N.: A Treatise on the Theory of Bessel Functions. Cambridge Univ. Press, 1922.

30. MCLACHLAN, NORMAN W.: Theory and Application of Mathieu Functions. Dover Publications, 1954.

31. TITCHMARSH, EDWARD C.: The Theory of Functions. Second ed., Oxford Univ. Press, 1939.

32. ERDELYI, A.: Asymptotic Expansions. Dover Publications, 1956.

33. LANGER, RUDOLPH E.: On the Asymptotic Solutions of Ordinary Differential Equations, with an Application to the Bessel Functions of Large Order. Trans. Am. Math. Soc., vol. 33, Jan. 1931, pp. 23–64.

34. LANGER, RUDOLPH E.: On the Asymptotic Solutions of Differential Equations, with an Application to the Bessel Functions of Large Complex Order. Trans. Am. Math. Soc., vol. 34, July 1932, pp. 447–480.

35. LANGER, R. E.: The Asymptotic Solution of Ordinary Linear Differential Equations of the Second Order, with Special Reference to the Stokes Phenomenon. Bull. Am. Math. Soc., vol. 40, 1934, pp. 545–582.

36. FRIEDRICHS, K. O.: Special Topics in Fluid Mechanics. New York Univ. Press, 1953, p. 126. Special Topics in Analysis. New York Univ. Press, 1954, p. 184.

37. LIGHTHILL, M. J.: A Technique for Rendering Approximate Solutions to Physical Problems Uniformly Valid. Phil. Mag., Ser. 7, vol. 40, no. 311, Dec. 1949, pp. 1179–1201.

38. KEVORKIAN, J.: The Two Variable Expansion Procedure for the Approximate Solution of Certain Non-Linear Differential Equations. Rep. SM–42620, Douglas Aircraft Co., Inc. (AD–437675), Dec. 3, 1962.

39. KOPAL, ZDENĚK: Numerical Analysis. John Wiley & Sons, Inc., 1955.

40. MILNE, WILLIAM E.: Numerical Solution of Differential Equations. John Wiley & Sons, Inc., 1953.

41. COLLATZ, L.: The Numerical Treatment of Differential Equations. Springer-Verlag, 1960.

42. FOX, L.: The Numerical Solution of Two-Point Boundary Problems in Ordinary Differential Equations. Clarendon Press, Oxford, 1957.

43. FOX, LESLIE; and MAYERS, D. F.: Computing Methods for Scientists and Engineers. Clarendon Press, Oxford, 1968.

44. HILDEBRAND, F. B.: Introduction to Numerical Analysis. McGraw-Hill Book Co., Inc., 1956.

45. CONTE, S. D.: Elementary Numerical Analysis—An Algorithmic Approach. McGraw-Hill Book Co., Inc., 1965.

46. RALSTON, ANTHONY; and WILF, HERBERT S., eds.: Mathematical Methods for Digital Computers. Vol. I. John Wiley & Sons, Inc., 1960.

47. VARGA, RICHARD S.: Matrix Iterative Analysis. Prentice-Hall, Inc., 1962.

48. BELLMAN, R.; and KALABA, R.: Quasilinearization and Nonlinear Boundary Valve Problems. American Elsevier, 1965.

INDEX

DIFFERENTIAL EQUATIONS

DIFFERENTIAL EQUATIONS